T0155991

SpringerBriefs in Probability and Mathematical Statistics

SpringerBriefs present concise summaries of cutting-edge research and practical applications across a wide spectrum of fields. Featuring compact volumes of 50 to 125 pages, the series covers a range of content from professional to academic. Briefs are characterized by fast, global electronic dissemination, standard publishing contracts, standardized manuscript preparation and formatting guidelines, and expedited production schedules.

Typical topics might include:

- A timely report of state-of-the art techniques
- A bridge between new research results, as published in journal articles, and a contextual literature review
- A snapshot of a hot or emerging topic
- Lecture of seminar notes making a specialist topic accessible for non-specialist readers
- SpringerBriefs in Probability and Mathematical Statistics showcase topics of current relevance in the field of probability and mathematical statistics

Manuscripts presenting new results in a classical field, new field, or an emerging topic, or bridges between new results and already published works, are encouraged. This series is intended for mathematicians and other scientists with interest in probability and mathematical statistics. All volumes published in this series undergo a thorough refereeing process.

The SBPMS series is published under the auspices of the Bernoulli Society for Mathematical Statistics and Probability.

More information about this series at http://www.springer.com/series/14353

Alfonso Rocha-Arteaga • Ken-iti Sato

Topics in Infinitely Divisible Distributions and Lévy Processes

Revised Edition

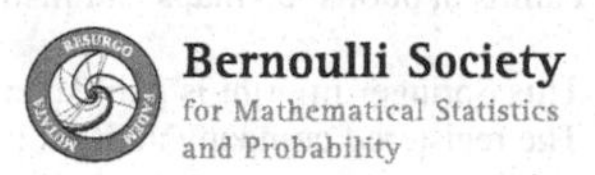

Alfonso Rocha-Arteaga
Facultad de Ciencias Físico-Matemáticas
Universidad Autónoma de Sinaloa
Culiacán, México

Ken-iti Sato
Hachiman-yama 1101-5-103
Tenpaku-ku, Nagoya, Japan

ISSN 2365-4333 ISSN 2365-4341 (electronic)
SpringerBriefs in Probability and Mathematical Statistics
ISBN 978-3-030-22699-2 ISBN 978-3-030-22700-5 (eBook)
https://doi.org/10.1007/978-3-030-22700-5

Mathematics Subject Classification: 60G51, 60E07, 60G18

This Springer imprint is published by the registered company Springer Nature Switzerland AG.
The registered company address is: Gewerbestrasse 11, 6330 Cham, Switzerland

Preface

Lévy processes are mathematical models of random phenomena that evolve in time and have applications in different branches of scientific interest and in modern areas, such as finance and risk theory. They are stochastic processes with independent and stationary increments in disjoint time intervals, continuous in probability, starting at zero, and with sample paths being right continuous with left limits. They are closely connected with infinitely divisible distributions, which have nth roots in convolution sense for each n. Infinitely divisible distributions are described by the Lévy–Khintchine representation of their Fourier transform (or characteristic function).

Selfdecomposable and stable distributions are major subclasses of the class of infinitely divisible distributions. The class of selfdecomposable distributions includes that of stable distributions, and furthermore, there is a decreasing chain of classes of distributions L_m, $m = 0, 1, \ldots, \infty$, from the class L_0 of selfdecomposable distributions to the class L_∞ generated by stable distributions through convolution and convergence.

This book deals with topics in the area of Lévy processes and infinitely divisible distributions, such as Ornstein–Uhlenbeck type processes, selfsimilar additive processes, and multivariate subordination; it focuses on developing them around the L_m classes. The prerequisites for this book are textbooks on probability theory on the level, such as Billingsley [13] (2012), Chung [19] (1974), Gnedenko and Kolmogorov [26] (1968), or Loève [58] (1977, 1978).

There are five chapters in this book. Chapter 1 studies the basic properties of L_m giving in particular two characterizations of them, the first one showing each member of L_m as a limit of distributions type equivalent of partial sums of a sequence of independent random variables and the second one as a special form of its Lévy–Khintchine representation. Chapter 2 introduces Ornstein–Uhlenbeck type processes generated by a Lévy process and a constant $c > 0$ through stochastic integrals based on Lévy processes. Necessary and sufficient conditions are given for a generating Lévy process so that the OU type process has a limit distribution of L_m class. A mapping from a class of distributions of the generating Lévy processes to the class of selfdecomposable distributions is defined; then, it is expressed by an improper stochastic integral. Its relationship with the L_m classes is studied. Chapter 3 establishes the correspondence between selfsimilar additive processes and selfdecomposable distributions and makes a close inspection of a transformation named after Lamperti, which makes c-selfsimilar additive processes and stationary OU type processes with parameter c transformed to each other. Chapters 4 and 5 treat multivariate subordination. Subordination is a procedure of combining two independent Lévy processes, one of which is increasing in $\mathbb{R}_+ = [0, \infty)$ and replaces the time parameter of the other. The resulting process is a new Lévy process. Chapter 4 generalizes it to multivariate subordination where one is a K-parameter Lévy process and the other is a K-valued Lévy process, K being a cone in $\mathbb{R}^N$. The relation of the Lévy–Khintchine representations involved is given when $K = \mathbb{R}_+^N$. Furthermore, the subordination of K-parameter convolution semigroups is shown for a general cone K. Chapter 5 studies the properties inherited by the subordinated process in multivariate subordination. Strictly stable and L_m properties are inherited by the subordinated from the subordinator when the subordinand is strictly stable.

This work began with Sato's visit to CIMAT, Guanajuato, in January–February 2001. It was published by Comunicaciones del CIMAT in 2001 and then by Sociedad Matemática Mexicana and Instituto de Matemáticas de la UNAM in 2003 in Aportaciones Matemáticas collection, Investigación series, number 17.

In this new edition, we have made a full revision of the previous publication. New material reflecting the advances in the understanding of the topics has been added. At the same time, many parts have been rewritten in an attempt to make them as close to self-contained as possible. Theorems, lemmas, propositions, and remarks were reorganized; some were deleted, and others were newly added. The pages on various extensions of L_m and $\mathfrak{S}_\alpha$ at the end of the previous edition were deleted; some of the comments there were moved to other places.

We thank Víctor Pérez-Abreu for his constant encouragement to us in the work for the old and new editions of this book. We also thank the anonymous reviewers for their valuable suggestions during our preparation for this edition.

Culiacán, México — Alfonso Rocha-Arteaga
Nagoya, Japan — Ken-iti Sato
April 2019

Contents

Chapter 1
Classes L_m and Their Characterization

Selfdecomposable distributions are extensions of stable distributions and form a subclass of the class of infinitely divisible distributions. In this chapter we will introduce, between the class L_0 of selfdecomposable distributions and the class $\mathfrak{S}$ of stable distributions, a chain of subclasses called $L_m, m = 1, \ldots, \infty$:

$$L_0 \supset L_1 \supset L_2 \supset \cdots \supset L_\infty \supset \mathfrak{S}.$$

In Sect. 1.1 basic properties are proved and the class L_m for $1 \le m < \infty$ is characterized as the class of limit distributions of partial sums of independent random variables whose distributions belong to L_{m-1}. The class L_∞ is defined as $\bigcap_{m<\infty} L_m$.

A representation of characteristic functions of the classes abovc is presented in Sect. 1.2, showing, in particular, that L_∞ is the smallest class containing the class $\mathfrak{S}$, closed under convolution and convergence. The representation of distributions in L_∞ indicates a clear connection to the representation of stable distributions. (The notation described at the end of thc book will bc frcely used.)

1.1 Basic Properties and Characterization by Limit Theorems

Let $\mathfrak{P} = \mathfrak{P}(\mathbb{R}^d)$ denote the class of probability measures (distributions) on the d-dimensional Euclidean space $\mathbb{R}^d$.

Definition 1.1 A distribution $\mu \in \mathfrak{P}$ is called *infinitely divisible* if, for each positive integer n, there is a $\mu_n \in \mathfrak{P}$ such that $\mu = \mu_n * \mu_n * \cdots * \mu_n$ (n factors), where $*$ denotes convolution. We write this equality as $\mu = \mu_n^{n*}$. The *class of infinitely divisible distributions* on $\mathbb{R}^d$ is denoted by $ID = ID(\mathbb{R}^d)$.

A. Rocha-Arteaga, K. Sato, *Topics in Infinitely Divisible Distributions and Lévy Processes*, SpringerBriefs in Probability and Mathematical Statistics,
https://doi.org/10.1007/978-3-030-22700-5_1

The *characteristic function* of $\mu \in \mathfrak{P}$ is defined by $\widehat{\mu}(z) = \int_{\mathbb{R}^d} e^{i\langle z,x\rangle}\mu(dx)$, $z \in \mathbb{R}^d$, where $\langle x, y\rangle = \sum_{j=1}^d x_j y_j$ for $x = (x_j)_{1\le j\le d}$ and $y = (y_j)_{1\le j\le d}$. It is continuous and satisfies $\widehat{\mu}(0) = 1$. If μ has a subscript, say μ_n, then we usually write $\widehat{\mu_n}$ as $\widehat{\mu}_n$. The characteristic function of a distribution uniquely determines the distribution. The characteristic function of the convolution of two distributions equals the product of their characteristic functions.

Throughout, *convergence in* $\mathfrak{P}$ is denoted by $\mu_n \to \mu$ and means *weak convergence of* μ_n to μ, that is, μ_n $(n = 1, 2, \ldots)$ and μ are in $\mathfrak{P}$ and $\int f(x)\mu_n(dx) \to \int f(x)\mu(dx)$ as $n \to \infty$ for all bounded continuous real-valued functions f on $\mathbb{R}^d$. If μ_n and μ are in $\mathfrak{P}$, then $\mu_n \to \mu$ and $\widehat{\mu}_n(z) \to \widehat{\mu}(z)$ for all $z \in \mathbb{R}^d$ are equivalent. In fact, $\mu_n \to \mu$ implies $\widehat{\mu}_n(z) \to \widehat{\mu}(z)$ uniformly on any compact set in $\mathbb{R}^d$. If μ_n, $n \in \mathbb{N}$, are in $\mathfrak{P}$ and satisfy $\widehat{\mu}_n(z) \to \varphi(z)$ for all $z \in \mathbb{R}^d$ for some function φ and if $\varphi(z)$ is continuous at $z = 0$, then φ is the characteristic function of some distribution.

Lemma 1.2 *Let φ be a continuous function from $\mathbb{R}^d$ into $\mathbb{C}$ satisfying $\varphi(0) = 1$ and $\varphi(z) \neq 0$ for all $z \in \mathbb{R}^d$. Then there is a unique continuous function f from $\mathbb{R}^d$ into $\mathbb{C}$ such that $f(0) = 0$ and $e^{f(z)} = \varphi(z)$ for $z \in \mathbb{R}^d$.*

See Theorem 7.6.2 of Chung [19] (1974) for $d = 1$ and Lemma 7.6 of [93] or [103] for general d.

Definition 1.3 Let φ and f be as in Lemma 1.2. Then, f is called the *distinguished logarithm* of φ and denoted by $\log \varphi$. For each $n \in \mathbb{N}$, $g_n(z) = e^{(1/n)(\log\varphi)(z)}$ is a unique continuous function $g_n \colon \mathbb{R}^d \to \mathbb{C}$ satisfying $g_n(z)^n = \varphi(z)$ and $g_n(0) = 1$. This function g_n is called the *distinguished nth root* of φ and denoted by $\varphi^{1/n}$. More generally we define $\varphi^t \colon \mathbb{R}^d \to \mathbb{C}$ by $\varphi^t(z) = e^{t(\log\varphi)(z)}$ for $t \in \mathbb{R}$ and call it the *distinguished tth power* of φ.

Keep in mind that $\log\varphi$ is not a composite function of φ and a fixed branch of the complex logarithmic function. Indeed, $(\log\varphi)(z_0)$ at $z_0 \in \mathbb{R}^d$ is determined not by $\varphi(z_0)$ but by $\{\varphi(tz_0)\colon 0 \le t \le 1\}$.

Proposition 1.4 *If $\mu \in ID$, then $\widehat{\mu}(z) \neq 0$ for all $z \in \mathbb{R}^d$.*

Proof of this fact is an application of the property of characteristic functions right above Lemma 1.2; see Lemma 7.5 of [93] or, for $d = 1$, Gnedenko and Kolmogorov [26] (1968), Chung [19] (1974), Loève [58] (1977, 1978).[1]

The converse of Proposition 1.4 is not true; for $d = 1$, if μ is the binomial distribution $b(n, p)$ with $p \neq 1/2$ or if $\mu = p\mu_{v_0} + (1-p)\mu_{v_1}$ with $0 < p < 1$ and $0 \le v_0 < v_1$, where μ_v is Gaussian distribution on $\mathbb{R}$ with mean 0 and variance v, then $\widehat{\mu}(z) \neq 0$ for all $z \in \mathbb{R}$ but $\mu \notin ID$ ([93] Corollary 24.4 and Remark 26.3).

[1] In Loève's book infinitely divisible distributions are named as infinitely decomposable.

It follows from Proposition 1.4 that, if $\mu \in ID$, then $\widehat{\mu}^t$ is definable for all $t \in \mathbb{R}$. It is the characteristic function of some distribution in ID for $t \in [0, \infty)$, as Proposition 1.5 will show.

A triplet $(\Omega, \mathcal{F}, P)$ is called a *probability space* if Ω is a set, $\mathcal{F}$ is a σ-algebra of subsets of Ω, and P is a probability measure defined on $\mathcal{F}$. When $(\Omega, \mathcal{F}, P)$ is a probability space, a mapping X from Ω into $\mathbb{R}^d$ such that $\{\omega \in \Omega \colon X(\omega) \in B\} \in \mathcal{F}$ for all $B \in \mathcal{B}(\mathbb{R}^d)$ is called a *random variable* on $\mathbb{R}^d$. If X is a random variable on $\mathbb{R}^d$, then $\mu(B) = P(\{\omega \in \Omega \colon X(\omega) \in B\})$ defines a probability measure on $\mathbb{R}^d$; this μ is called the *distribution* (*law*) of X and denoted by $\mathcal{L}(X)$. Sometimes we write $\mu_X = \mathcal{L}(X)$. A collection $\{X_t \colon t \geq 0\}$ of random variables on $\mathbb{R}^d$ defined on a common probability space is called a *stochastic process*. It is denoted by $\{X_t\}$. Two stochastic processes are said to be *identical in law* if they have a common system of finite-dimensional distributions. We use the word stochastic process also when the parameter set is other than $[0, \infty)$.

The class ID has the following properties.

Proposition 1.5

(i) If μ_1 and μ_2 are in ID, then $\mu_1 * \mu_2 \in ID$ and $\log \widehat{\mu_1 * \mu_2} = \log \widehat{\mu}_1 + \log \widehat{\mu}_2$.
(ii) If $\mu_n \in ID$, $\mu \in \mathfrak{P}$, and $\mu_n \to \mu$, then $\mu \in ID$.
(iii) If $\mu_1 = \mathcal{L}(X) \in ID$ and $\mu_2 = \mathcal{L}(aX + c)$ with $a \in \mathbb{R}$ and $c \in \mathbb{R}^d$, then $\mu_2 \in ID$.
(iv) If $\mu \in ID$, then, for any $t \in [0, \infty)$, $\widehat{\mu}^t$ is the characteristic function of a distribution in ID.

Proof Let us show (i). Since convolution is commutative and associative, it follows from $\mu_1 = (\mu_{1,k})^{k*}$ and $\mu_2 = (\mu_{2,k})^{k*}$ for $k \in \mathbb{N}$ that $\mu_1 * \mu_2 = (\mu_{1,k} * \mu_{2,k})^{k*}$. Hence, if $\mu_1, \mu_2 \in ID$, then $\mu_1 * \mu_2 \in ID$ and moreover $\log \widehat{\mu_1 * \mu_2} = \log \widehat{\mu}_1 + \log \widehat{\mu}_2$ since the right-hand side is continuous and vanishes at 0 and

$$e^{\log \widehat{\mu}_1 + \log \widehat{\mu}_2} = e^{\log \widehat{\mu}_1} e^{\log \widehat{\mu}_2} = \widehat{\mu}_1 \widehat{\mu}_2 = \widehat{\mu_1 * \mu_2}.$$

For the proof of (ii), note that in general $\mu_n \to \mu$ in $\mathfrak{P}$ implies $\widehat{\mu}_n(z) \to \widehat{\mu}(z)$ uniformly on any compact set in $\mathbb{R}^d$ and see the proof of Lemma 7.8 of [93] or, for $d = 1$, see [19, 26, 58]. To show (iii), it is enough to notice that $\widehat{\mu}_2(z) = \widehat{\mu}_1(az) e^{i\langle c,z\rangle}$. To show (iv), let $\mu \in ID$. Then, for each $k \in \mathbb{N}$, there is $\mu_k \in \mathfrak{P}$ such that $\mu = \mu_k^{k*}$. Since $\widehat{\mu}(z) = (\widehat{\mu}_k(z))^k$ and $\widehat{\mu}_k$ is continuous with $\widehat{\mu}_k(0) = 1$, we have $\widehat{\mu}_k = \widehat{\mu}^{1/k}$. Hence μ_k is uniquely determined by μ and k. It follows that $\widehat{\mu}_{k_1 k_2} = (\widehat{\mu}_{k_2})^{1/k_1}$ for $k_1, k_2 \in \mathbb{N}$ and that $\mu_k \in ID$ for each k. Now approximate $t \in [0, \infty)$ by positive rationals and use (i) and (ii) to conclude that $\widehat{\mu}^t$ is the characteristic function of a distribution in ID. ■

Definition 1.6 Let $\mu \in ID$. The distribution with characteristic function $\widehat{\mu}^t$ in Proposition 1.5 (iv) is denoted by μ^{t*} or, sometimes, by μ^t.

Remark 1.7 We have $\widehat{\mu^{t*}}(z) = e^{t(\log \widehat{\mu})(z)}$, since $\widehat{\mu}^t$ is defined as in Definition 1.3. Hence $\widehat{\mu^{n*}}(z) = \left(e^{\log \widehat{\mu}(z)}\right)^n = \widehat{\mu}(z)^n$ for $n \in \mathbb{N}$, which means that μ^{n*} is the same as that in Definition 1.1. Note that $\mu^{0*} = \delta_0$ since $\widehat{\mu^{0*}} = 1$. Also note that $(\mu_1 * \mu_2)^{t*} = \mu_1^{t*} * \mu_2^{t*}$, since, for $\rho = (\mu_1 * \mu_2)^{t*}$, $\widehat{\rho} = e^{t \log \widehat{\mu_1 * \mu_2}} = e^{t(\log \widehat{\mu}_1 + \log \widehat{\mu}_2)} = e^{t \log \widehat{\mu}_1} e^{t \log \widehat{\mu}_2} = \widehat{\mu_1^{t*} \mu_2^{t*}}$.

Let us introduce the notion of a null array.

Definition 1.8 A double sequence of random variables $\{Z_{nk} : k = 1, 2, \ldots, r_n;\ n \in \mathbb{N}\}$ on $\mathbb{R}^d$ is called a *null array* on $\mathbb{R}^d$ if, for each fixed n, $Z_{n1}, Z_{n2}, \ldots, Z_{nr_n}$ are independent and if, for every $\varepsilon > 0$,

$$\lim_{n\to\infty} \max_{1\le k\le r_n} P[\,|Z_{nk}| > \varepsilon\,] = 0. \tag{1.1}$$

The sums $S_n = \sum_{k=1}^{r_n} Z_{nk}$, $n = 1, 2, \ldots$, are called the *row sums*.

Remark 1.9 In Definition 1.8 the condition (1.1) can be replaced by

$$\lim_{n\to\infty} \max_{1\le k\le r_n} |\widehat{\mu}_{Z_{nk}}(z) - 1| = 0, \qquad z \in \mathbb{R}^d. \tag{1.2}$$

Indeed, to see that (1.2) implies (1.1) for $d = 1$, see that, for each $\mu \in \mathfrak{P}(\mathbb{R})$ and $a > 0$,

$$\begin{aligned} a^{-1} \textstyle\int_{-a}^{a} (1 - \widehat{\mu}(z))dz &= a^{-1} \textstyle\int \mu(dx) \textstyle\int_{-a}^{a} (1 - e^{izx})dz \\ &= 2\textstyle\int (1 - (\sin ax)/(ax))\mu(dx) \ge 2 \textstyle\int_{|x|\ge 2/a} (1 - (a|x|)^{-1})\mu(dx) \\ &\ge \mu(\{x\colon |x| \ge 2/a\}). \end{aligned}$$

The case of general d is reduced to the case $d = 1$ since, for $X = (X_j)_{1\le j\le d}$, $P[\,|X| > \varepsilon\,] \le \sum_{j=1}^{d} P[\,|X_j| > \varepsilon/\sqrt{d}\,]$ and $\widehat{\mu}_{X_j}(z)$ is the value of $\widehat{\mu}_X$ on the jth axis. To see that (1.1) implies (1.2), notice that, for $\mu \in \mathfrak{P}(\mathbb{R}^d)$,

$$|\widehat{\mu}(z) - 1| \le \textstyle\int |e^{i\langle z,x\rangle} - 1|\, \mu(dx) \le 2 \textstyle\int_{|x|>\varepsilon} \mu(dx) + |z|\varepsilon$$

for every $\varepsilon > 0$.

The following theorem is one of the fundamental results on infinitely divisible distributions.

Theorem 1.10 *Let $\{Z_{nk}\}$ be a null array on $\mathbb{R}^d$ with row sums S_n. If, for some $c_n \in \mathbb{R}^d$, $n = 1, 2, \ldots$, the distribution of $S_n - c_n$ converges to some $\mu \in \mathfrak{P}$, then $\mu \in ID$.*

This is shown in Theorem 9.3 of [93] or, for $d = 1$, in [19, 26, 58].[2] Conversely, any $\mu \in ID$ is the limit of the distribution of the row sum of a null array, as is seen from the definition of infinite divisibility.

Now we introduce selfdecomposable distributions.

Definition 1.11 A distribution $\mu \in \mathfrak{P}$ is *selfdecomposable* if, for every $b \in (1, \infty)$ there is $\rho_b \in \mathfrak{P}$ such that

$$\widehat{\mu}(z) = \widehat{\mu}\left(b^{-1}z\right)\widehat{\rho_b}(z). \tag{1.3}$$

Sometimes it is called of *class L*. Let $L_0 = L_0\left(\mathbb{R}^d\right)$ be the *class of selfdecomposable distributions* on $\mathbb{R}^d$.

Gaussian distributions on $\mathbb{R}^d$ and Γ-distributions and Cauchy distributions on $\mathbb{R}$ are examples of selfdecomposable distributions.

In order to analyse L_0, let us use the following operation $\mathcal{K}$ that makes a subclass of $\mathfrak{P}$ from a subclass of $\mathfrak{P}$.

Definition 1.12 For any subclass $\mathfrak{Q}$ of $\mathfrak{P}$, define a subclass $\mathcal{K}(\mathfrak{Q})$ of $\mathfrak{P}$ as follows: $\mathcal{K}(\mathfrak{Q})$ is the totality of $\mu \in \mathfrak{P}$ such that there are independent random variables $Z_1, Z_2, \ldots$ on $\mathbb{R}^d$, $b_n > 0$, and $c_n \in \mathbb{R}^d$ satisfying

(a) $\mathcal{L}\left(b_n \sum_{k=1}^{n} Z_k - c_n\right) \to \mu$ as $n \to \infty$,
(b) $\{b_n Z_k : k = 1, 2, \ldots, n; n \in \mathbb{N}\}$ is a null array,
(c) $\mathcal{L}(Z_k) \in \mathfrak{Q}$ for each k.

It follows from Theorem 1.10 that $\mathcal{K}(\mathfrak{Q}) \subset ID$.

Proposition 1.13 *Let $\mu \in L_0$. Then ρ_b in (1.3) is uniquely determined by μ and b, and both μ and ρ_b are in ID.*

Proof Let $\mu \in L_0$. Then $\widehat{\mu}$ has no zero. Indeed, if it has a zero, then there is z_0 such that $\widehat{\mu}(z_0) = 0$ and $\widehat{\mu}(z) \neq 0$ for $|z| < |z_0|$ and hence we have $\widehat{\rho_b}(z_0) = 0$ for all $b > 1$ from (1.3) and

$$1 = \mathrm{Re}(1 - \widehat{\rho_b}(z_0)) \leq 4\mathrm{Re}(1 - \widehat{\rho_b}(2^{-1}z_0)) = 4\mathrm{Re}\left(1 - \frac{\widehat{\mu}(2^{-1}z_0))}{\widehat{\mu}(2^{-1}b^{-1}z_0)}\right) \to 0$$

as $b \downarrow 1$, which is absurd. Here we have used the fact that, for any $\mu \in \mathfrak{P}$,

$$\begin{aligned}\mathrm{Re}(1 - \widehat{\mu}(2z)) &= \textstyle\int(1 - \cos\langle 2z, x\rangle)\mu(dx) = 2\int(1 - \cos^2\langle z, x\rangle)\mu(dx)\\ &\leq 4\textstyle\int(1 - \cos\langle z, x\rangle)\mu(dx) = 4\mathrm{Re}(1 - \widehat{\mu}(z)).\end{aligned}$$

[2] The property (1.1) for Z_{nk} is called *infinitesimal* by Gnedenko and Kolmogorov [26] (1968) and *uniformly asymptotically negligible* by Loève [58] (1977, 1978). Chung [19] (1974) uses the word *holospoudic double array*. We follow Feller [24] (1971) in using the word "null array".

It follows from $\widehat{\mu}(z) \neq 0$ for all z that $\widehat{\rho_b}$ (and hence ρ_b itself) in (1.3) is uniquely determined by μ and b. Now let us see that μ and ρ_b are in ID. Let $Z_1, Z_2, \ldots$ be independent random variables with $\widehat{\mu}_{Z_k}(z) = \widehat{\rho}_{(k+1)/k}((k+1)z) = \widehat{\mu}((k+1)z)/\widehat{\mu}(kz)$. Then $S_n = \sum_{k=1}^{n} Z_k$ satisfies

$$E[e^{i\langle z, n^{-1}S_n\rangle}] = \prod_{k=1}^{n} \frac{\widehat{\mu}((k+1)n^{-1}z)}{\widehat{\mu}(kn^{-1}z)} = \frac{\widehat{\mu}((n+1)n^{-1}z)}{\widehat{\mu}(n^{-1}z)} \to \widehat{\mu}(z)$$

as $n \to \infty$. We see that $\{n^{-1}Z_k \colon k = 1, \ldots, n;\ n \in \mathbb{N}\}$ is a null array, since

$$\max_{1 \le k \le n} \left| \frac{\widehat{\mu}((k+1)n^{-1}z)}{\widehat{\mu}(kn^{-1}z)} - 1 \right| = \max_{1 \le k \le n} \frac{\left|\widehat{\mu}((k+1)n^{-1}z) - \widehat{\mu}(kn^{-1}z)\right|}{\left|\widehat{\mu}(kn^{-1}z)\right|} \to 0,$$

which is (1.2). Hence $\mu \in \mathcal{K}(\mathfrak{P}) \subset ID$ with $b_n = n^{-1}$ and $c_n = 0$. In order to prove $\rho_b \in ID$, let m_j and n_j be positive integers such that $m_j < n_j$, $n_j/m_j \to b$, and $m_j \to \infty$ as $j \to \infty$. Let

$$W_n = n^{-1}S_n, \quad U_j = n_j^{-1}\sum_{k=1}^{m_j} Z_k, \quad V_j = n_j^{-1}\sum_{k=m_j+1}^{n_j} Z_k.$$

Then $\mu_{W_n} \to \mu$, $\widehat{\mu}_{W_{n_j}} = \widehat{\mu}_{U_j}\widehat{\mu}_{V_j}$, and $U_j = m_j n_j^{-1} W_{m_j}$. Thus

$$\begin{aligned}\left|\widehat{\mu}_{U_j}(z) - \widehat{\mu}(m_j n_j^{-1}z)\right| &= \left|\widehat{\mu}_{W_{m_j}}(m_j n_j^{-1}z) - \widehat{\mu}(m_j n_j^{-1}z)\right| \\ &\le \max_{|w| \le |z|} \left|\widehat{\mu}_{W_{m_j}}(w) - \widehat{\mu}(w)\right| \to 0\end{aligned}$$

as $j \to \infty$ and hence $\widehat{\mu}_{U_j}(z) \to \widehat{\mu}(b^{-1}z)$. It follows that $\widehat{\mu}_{V_j}(z) \to \widehat{\rho_b}(z)$ for each z. Since $\left\{n_j^{-1}Z_k \colon k = m_j + 1, \ldots, n_j;\ j \in \mathbb{N}\right\}$ is a null array, we see that $\rho_b \in ID$, using Theorem 1.10. ■

Definition 1.14 For $m = 1, 2, 3, \ldots$, $L_m = L_m\left(\mathbb{R}^d\right)$ is recursively defined as follows: $\mu \in L_m$ if and only if for every $b > 1$ there is $\rho_b \in L_{m-1}$ such that $\widehat{\mu}(z) = \widehat{\mu}\left(b^{-1}z\right)\widehat{\rho_b}(z)$. Sometimes $\mu \in L_m$ is called $m+1$ *times selfdecomposable* in view of Definition 1.11 and this naming is made more precise in Theorem 2.29 proved later.

It is immediate that $L_0 \supset L_m$ for all $m \ge 1$. Next, we prove that these classes form a nested sequence. Thus, intersection over all L_m will give the limiting class.

Proposition 1.15 $ID \supset L_0 \supset L_1 \supset L_2 \supset \cdots$.

Proof We already have $ID \supset L_0 \supset L_1$. Suppose that $L_m \supset L_{m+1}$. Then $L_{m+1} \supset L_{m+2}$ follows from the definition. ■

Definition 1.16 $L_\infty = L_\infty(\mathbb{R}^d) = \bigcap_{m=0}^{\infty} L_m(\mathbb{R}^d)$.

Remark 1.17 Trivial distributions δ_c on $\mathbb{R}^d$ belong to L_∞, since $\widehat{\delta_c}(z) = \widehat{\delta_c}(b^{-1}z)\,\widehat{\delta}_{(1-b^{-1})c}(z)$ for all $b > 1$.

The classes L_m have the following properties.

Proposition 1.18 *Let $m \in \{0, 1, 2, \ldots, \infty\}$.*

(i) If μ_1 and μ_2 are in L_m, then $\mu_1 * \mu_2 \in L_m$.
(ii) If $\mu_n \in L_m$ and $\mu_n \to \mu$, then $\mu \in L_m$.
(iii) If $\mu_1 = \mathcal{L}(X) \in L_m$ and $\mu_2 = \mathcal{L}(aX + c)$ with $a \in \mathbb{R}$ and $c \in \mathbb{R}^d$, then $\mu_2 \in L_m$.
(iv) If $\mu \in L_m$, then, for any $t \in [0, \infty)$, $\mu^{t*} \in L_m$.

Proof Let us define $L_{-1} = ID$ and prove (i)–(iv) for $m = -1, 0, 1, 2, \ldots$ by induction. The assertions (i)–(iv) for L_{-1} are shown in Proposition 1.5. Now we assume that, for a given $m \geq 0$, (i)–(iv) are true with m replaced by $m - 1$ and show (i)–(iv) for L_m. For μ_j in L_0, the corresponding ρ_b in (1.3) is denoted by $\rho_{j,b}$.

(i) If $\mu_1, \mu_2 \in L_m$, then $\rho_{1,b}, \rho_{2,b} \in L_{m-1}$ and ρ_b for $\mu_1 * \mu_2$ equals $\rho_{1,b} * \rho_{2,b}$, which is in L_{m-1}.

(ii) If $\mu_n \in L_m$ and $\mu_n \to \mu$, then $\rho_{n,b} \in L_{m-1}$ and

$$\widehat{\rho}_{n,b}(z) = \widehat{\mu}_n(z) / \widehat{\mu}_n\left(b^{-1}z\right) \to \widehat{\mu}(z)/\widehat{\mu}\left(b^{-1}z\right),$$

which is continuous in z and is the characteristic function of some $\rho_b \in \mathfrak{P}$. The induction hypothesis implies $\rho_b \in L_{m-1}$. It follows that $\mu \in L_m$.

(iii) Let $\mu_1 = \mathcal{L}(X) \in L_m$ and $\mu_2 = \mathcal{L}(aX + c)$. Then $\rho_{1,b} \in L_{m-1}$ and

$$\begin{aligned}
\widehat{\mu}_2(z) = \widehat{\mu}_1(az)\, e^{i\langle c,z\rangle} &= \widehat{\mu}_1\left(b^{-1}az\right) \widehat{\rho}_{1,b}(az)\, e^{i\langle c,z\rangle} \\
&= \widehat{\mu}_1\left(b^{-1}az\right) e^{i\langle c,b^{-1}z\rangle} e^{i\langle (1-b^{-1})c,z\rangle} \widehat{\rho}_{1,b}(az) \\
&= \widehat{\mu}_2\left(b^{-1}z\right) \widehat{\delta}_{(1-b^{-1})c}(z)\, \widehat{\rho}_{1,b}(az).
\end{aligned}$$

We see $\mu_2 \in L_m$ since $\widehat{\delta}_{(1-b^{-1})c}(z)\,\widehat{\rho}_{1,b}(az)$ is the characteristic function of a distribution in L_{m-1}.

(iv) Let $\mu \in L_m$. Then $\rho_b \in L_{m-1}$. It follows from (1.3) that $(\log\widehat{\mu})(z) = (\log\widehat{\mu})(b^{-1}z) + (\log\widehat{\rho}_b)(z)$. Hence $\widehat{\mu}^t(z) = \widehat{\mu}^t(b^{-1}z)\widehat{\rho}_b^t(z)$. Since $\rho_b^t \in L_{m-1}$, we obtain $\mu^{t*} \in L_m$.

The assertions (i)–(iv) for $m = \infty$ are consequences of the assertions for $m < \infty$. ∎

Next we introduce stable distributions and prove that they are contained in the class L_∞. In the next section it will be shown that L_∞ is the smallest class closed under convolution and convergence that contains the stable distributions.

Definition 1.19 A distribution $\mu \in \mathfrak{P}$ is called *stable* if, for each $n \in \mathbb{N}$, there are $a_n \in (0, \infty)$ and $c_n \in \mathbb{R}^d$ satisfying

$$\widehat{\mu}(z)^n = \widehat{\mu}(a_n z) e^{i\langle c_n, z\rangle}, \qquad z \in \mathbb{R}^d. \tag{1.4}$$

A distribution $\mu \in \mathfrak{P}$ is called *strictly stable* if μ is stable and (1.4) holds with $c_n = 0$. The class of stable or strictly stable distributions, respectively, on $\mathbb{R}^d$ is denoted by $\mathfrak{S} = \mathfrak{S}(\mathbb{R}^d)$ or $\mathfrak{S}^0 = \mathfrak{S}^0(\mathbb{R}^d)$.

An underlying concept in this definition is type equivalence. Two distributions $\mathcal{L}(X)$ and $\mathcal{L}(Y)$ are said to be *type equivalent* if $\mathcal{L}(Y) = \mathcal{L}(aX + c)$ for some $a > 0$ and $c \in \mathbb{R}^d$.

Remark 1.20 If $\mu \in \mathfrak{S}$, then $\mu \in ID$, because

$$\widehat{\mu}(z) = \left\{ \widehat{\mu}(a_n^{-1} z) e^{-i\langle a_n^{-1} n^{-1} c_n, z\rangle} \right\}^n.$$

If $\mu \in \mathfrak{S}$, then $\mu^{s*} \in \mathfrak{S}$ for all $s \geq 0$. Indeed, rewrite (1.4) as $n(\log \widehat{\mu})(z) = (\log \widehat{\mu})(a_n z) + i\langle c_n, z\rangle$ and then, multiplying by s, obtain $n(\log \widehat{\mu^s})(z) = (\log \widehat{\mu^s})(a_n z) + i\langle s c_n, z\rangle$. Similarly, if $\mu \in \mathfrak{S}^0$, then $\mu^{s*} \in \mathfrak{S}^0$ for all $s \geq 0$.

For stable distributions the following Propositions 1.21 and 1.22 hold. Their detailed proofs are found in Section 13 and solution of exercise E18.4 of [93].[3]

Proposition 1.21 *A distribution $\mu \in \mathfrak{P}$ is in $\mathfrak{S}$ if and only if $\mu \in ID$ and, for every $t \in (0, \infty)$, there exist $a_t \in (0, \infty)$ and $c_t \in \mathbb{R}^d$ such that*

$$\widehat{\mu}^t(z) = \widehat{\mu}(a_t z) e^{i\langle c_t, z\rangle}, \qquad z \in \mathbb{R}^d. \tag{1.5}$$

A distribution $\mu \in \mathfrak{P}$ is in $\mathfrak{S}^0$ if and only if $\mu \in ID$ and, for every $t \in (0, \infty)$, there is $a_t \in (0, \infty)$ such that

$$\widehat{\mu}^t(z) = \widehat{\mu}(a_t z), \qquad z \in \mathbb{R}^d. \tag{1.6}$$

Proposition 1.22 *If $\mu \in \mathfrak{S}$ and μ is not trivial, then a_t and c_t in (1.5) are unique for every $t \in (0, \infty)$ and there is $\alpha \in (0, 2]$ such that, for all $t \in (0, \infty)$, $a_t = t^{1/\alpha}$. If $\mu \in \mathfrak{S}^0$ and $\mu \neq \delta_0$, then a_t in (1.6) is unique for every $t \in (0, \infty)$ and there is $\alpha \in (0, 2]$ such that, for all $t \in (0, \infty)$, $a_t = t^{1/\alpha}$.*

---

[3] The proof of Proposition 1.22 uses Lévy–Khintchine representation in Theorem 1.28. But Theorem 1.28 is shown independently of the theory of stable distributions.

Definition 1.23[4] Let $\alpha \in (0, 2]$. A distribution $\mu \in \mathfrak{P}(\mathbb{R}^d)$ is called *α-stable* or stable with *index α* if $\mu \in ID$ and, for every $t \in (0, \infty)$, there is $c_t \in \mathbb{R}^d$ such that

$$\widehat{\mu}^t(z) = \widehat{\mu}(t^{1/\alpha} z) e^{i\langle c_t, z\rangle}. \tag{1.7}$$

A distribution $\mu \in \mathfrak{P}(\mathbb{R}^d)$ is called *strictly α-stable* or strictly stable with *index α* if $\mu \in ID$ and, for every $t \in (0, \infty)$,

$$\widehat{\mu}^t(z) = \widehat{\mu}(t^{1/\alpha} z). \tag{1.8}$$

The class of α-stable or strictly α-stable distributions, respectively, on $\mathbb{R}^d$ is denoted by $\mathfrak{S}_\alpha = \mathfrak{S}_\alpha(\mathbb{R}^d)$ or $\mathfrak{S}^0_\alpha = \mathfrak{S}^0_\alpha(\mathbb{R}^d)$.

A distribution $\mu \in \mathfrak{P}(\mathbb{R}^d)$ is called *degenerate* if there are $c \in \mathbb{R}^d$ and a linear subspace V of $\mathbb{R}^d$ with $\dim(V) \le d - 1$ such that $\mu(c + V) = 1$. Otherwise μ is called *nondegenerate*. The class $\mathfrak{S}_2$ is the class of (possibly degenerate) Gaussian distributions.

Proposition 1.24 *If $\alpha, \alpha' \in (0, 2]$ and $\alpha \neq \alpha'$, then*

$$\mathfrak{S}_\alpha \cap \mathfrak{S}_{\alpha'} = \{\delta_c \colon c \in \mathbb{R}^d\} \quad \text{and} \quad \mathfrak{S}^0_\alpha \cap \mathfrak{S}^0_{\alpha'} = \{\delta_0\}. \tag{1.9}$$

Any trivial distribution is strictly 1-stable. For every $\alpha \neq 1$ and $c \neq 0$, we have $\delta_c \notin \mathfrak{S}^0_\alpha$. We have

$$\mathfrak{S} = \bigcup_{\alpha \in (0,2]} \mathfrak{S}^0_\alpha \quad \text{and} \quad \mathfrak{S}^0 = \bigcup_{\alpha \in (0,2]} \mathfrak{S}^0_\alpha. \tag{1.10}$$

If $\mu \in \mathfrak{S}_\alpha$, then $\mu^{s*} \in \mathfrak{S}_\alpha$ for all $s \ge 0$. If $\mu \in \mathfrak{S}^0_\alpha$, then $\mu^{s*} \in \mathfrak{S}^0_\alpha$ for all $s \ge 0$.

Proof This follows from Proposition 1.22 and Definition 1.23. The assertion on μ^{s*} is proved similarly to Remark 1.20. ■

Proposition 1.25 $L_\infty \supset \mathfrak{S}$.

Proof Let $\mu \in \mathfrak{S}$. Then, for some $\alpha \in (0, 2]$, (1.7) holds. Given $b > 1$ let $t = b^{-\alpha} < 1$. Then

$$\widehat{\mu}(z) = \widehat{\mu}(z)^{1-t} \widehat{\mu}\left(b^{-1} z\right) e^{i\langle c, z\rangle} = \widehat{\mu}\left(b^{-1} z\right) \widehat{\rho_b}(z),$$

where ρ_b is given by $\widehat{\rho_b}(z) = \widehat{\mu}(z)^{1-t} e^{i\langle c, z\rangle}$. Hence $\mu \in L_0$. Now, by Proposition 1.18 and Remark 1.17, $\rho_b \in L_0$. The last arguments are recursively applied to yield

[4] This definition of α-stability and strict α-stability is slightly different from that of [93], where trivial distributions do not have index in stability and δ_0 does not have index in strict stability. Thus neither (1.9) nor (1.10) is true if $\mathfrak{S}_\alpha$ and $\mathfrak{S}^0_\alpha$ are, respectively, the classes of α-stable and strictly α-stable distributions in the sense of [93].

$$\rho_b \in L_0 \Rightarrow \mu \in L_1 \Rightarrow \rho_b \in L_1 \Rightarrow \mu \in L_2 \Rightarrow \rho_b \in L_2 \Rightarrow \mu \in L_3 \Rightarrow \dots$$

Thus $\mu \in L_m$ for every $m \geq 0$. ■

Using the operation $\mathcal{K}$ of Definition 1.12, we characterize the classes L_m for $m = 0, 1, 2, \dots, \infty$ as classes of limit distributions. This is a main theorem in this section.

Theorem 1.26

(i) $L_0 = \mathcal{K}(\mathfrak{P}) = \mathcal{K}(ID)$.
(ii) $L_m = \mathcal{K}(L_{m-1})$ *for* $m = 1, 2, \dots$
(iii) $L_\infty = \mathcal{K}(L_\infty)$ *and* L_∞ *is the greatest class* $\mathfrak{Q}$ *that satisfies* $\mathfrak{Q} = \mathcal{K}(\mathfrak{Q})$.

Proof

(i) Proposition 1.13 and its proof show that $L_0 \subset \mathcal{K}(ID) \subset \mathcal{K}(\mathfrak{P})$. Hence it is enough to show that $\mathcal{K}(\mathfrak{P}) \subset L_0$. Suppose that $\mu \in \mathcal{K}(\mathfrak{P})$ with Z_k, b_n, and c_n being those of Definition 1.12. Then $\mu \in ID$. If μ is trivial, then $\mu \in L_0$ as is shown by Remark 1.17. So we assume that μ is non-trivial. Then we have $b_n \to 0$ and $b_n/b_{n+1} \to 1$ as $n \to \infty$; see Lemma 15.4 of [93] for details. Thus $\log b_n \to -\infty$ and $\log b_n - \log b_{n+1} \to 0$. Then for every $b > 1$ there are sequences of positive integers $\{m_j\}$, $\{n_j\}$ going to infinity such that $m_j < n_j$ and $\log b_{m_j} - \log b_{n_j} \to \log b$, that is, $b_{m_j}/b_{n_j} \to b$. Let

$$W_n = b_n \sum_{k=1}^{n} Z_k + c_n,$$

$$U_j = b_{n_j} \sum_{k=1}^{m_j} Z_k + b_{n_j} b_{m_j}^{-1} c_{m_j}, \quad V_j = b_{n_j} \sum_{k=m_j+1}^{n_j} Z_k + c_{n_j} - b_{n_j} b_{m_j}^{-1} c_{m_j}.$$

Then $\mu_{W_n} \to \mu$, $W_{n_j} = U_j + V_j$, and

$$\widehat{\mu}_{W_{n_j}}(z) = \widehat{\mu}_{U_j}(z)\, \widehat{\mu}_{V_j}(z) \tag{1.11}$$

by independence of $\{Z_k\}$. Since $U_j = b_{n_j} b_{m_j}^{-1} W_{m_j}$, we have $\widehat{\mu}_{U_j}(z) \to \widehat{\mu}\left(b^{-1}z\right)$ as $j \to \infty$, as in the last part of the proof of Proposition 1.13. Now it follows from (1.11) that, for each z, $\widehat{\mu}_{V_j}(z)$ tends to $\widehat{\mu}(z)/\widehat{\mu}\left(b^{-1}z\right)$, which is continuous in z. Hence there is $\rho_b \in \mathfrak{P}$ such that $\widehat{\rho_b}(z) = \widehat{\mu}(z)/\widehat{\mu}\left(b^{-1}z\right)$. Therefore $\mu \in L_0$.

(ii) Let m be a positive integer. Let $\mu \in L_m$. Then ρ_b in (1.3) is in L_{m-1}. Let us show that $\mu \in \mathcal{K}(L_{m-1})$. Since $\mu \in ID$, $\widehat{\mu}$ has no zero and $\widehat{\rho_b}(z) = \widehat{\mu}(z)/\widehat{\mu}\left(b^{-1}z\right)$ for $b > 1$. Let $\{Z_k\}$ be independent random variables with $\widehat{\mu}_{Z_k}(z) = \widehat{\rho}_{(k+1)/k}((k+1)z)$. Then $\mathcal{L}(Z_k) \in L_{m-1}$ by Proposition 1.18 (iii). The same argument as in the proof of Proposition 1.13 now shows that $\mu \in$

$\mathcal{K}(L_{m-1})$ with $b_n = n^{-1}$ and $c_n = 0$ in Definition 1.12. The converse part in (ii) saying that $\mathcal{K}(L_{m-1}) \subset L_m$ is proved almost in the same way as the proof of $\mathcal{K}(\mathfrak{P}) \subset L_0$ in (i); simply note that $\mathcal{L}(Z_k) \in L_{m-1}$ implies that $\mathcal{L}(V_j) \in L_{m-1}$ and $\rho_b \in L_{m-1}$ by the use of Proposition 1.18.

(iii) We have $L_m = \mathcal{K}(L_{m-1}) \supset \mathcal{K}(L_\infty)$ for all m and, in consequence, $L_\infty \supset \mathcal{K}(L_\infty)$. Conversely, if $\mu \in L_\infty$, then $\mu \in \mathcal{K}(L_\infty)$. Indeed, we have $\rho_b \in L_\infty$. Let us take Z_k as in the proof of the converse part of (ii) and see that $\mathcal{L}\left(\sum_{k=1}^n Z_k/n\right)$ tends to μ, where the distribution of Z_k is in L_∞ and $\{Z_k/n\}$ is a null array. This means that $\mu \in \mathcal{K}(L_\infty)$. Hence $L_\infty = \mathcal{K}(L_\infty)$.

Finally, let $\mathfrak{Q} = \mathcal{K}(\mathfrak{Q})$. Then $\mathfrak{Q} \subset \mathcal{K}(\mathfrak{P}) = L_0$. Now we obtain $\mathfrak{Q} = \mathcal{K}(\mathfrak{Q}) \subset \mathcal{K}(L_0) = L_1$. Repetition of this argument using (ii) yields $\mathfrak{Q} \subset L_m$ for every m. Therefore $\mathfrak{Q} \subset L_\infty$. ■

Characterization of $\mathfrak{S}$ as a class of limit distributions is as follows.

Theorem 1.27 *A distribution $\mu \in \mathfrak{P}$ is stable if and only if $\mu \in \mathcal{K}(\mathfrak{Q})$ for some $\mathfrak{Q}$ consisting of one element. In this case $\{Z_k\}$ in Definition 1.12 are independent identically distributed random variables on $\mathbb{R}^d$ and $\mathcal{K}(\mathfrak{Q})$ is the class of distributions type equivalent with μ.*

The "only if" part of this theorem is a direct consequence of the definition of stability. For the "if" part, see Theorem 15.7 of [93] or, for $d = 1$, [24, 26, 58].

1.2 Characterization in Lévy–Khintchine Representation

One of the basic results on infinitely divisible distributions is the following *Lévy–Khintchine representation* of their characteristic functions ([93] Theorem 8.1).

Theorem 1.28 *If $\mu \in ID(\mathbb{R}^d)$, then*

$$\widehat{\mu}(z) = \exp\left[-\tfrac{1}{2}\langle z, Az\rangle + i\langle \gamma, z\rangle + \int_{\mathbb{R}^d}\left(e^{i\langle z,x\rangle} - 1 - i\langle z, x\rangle 1_{\{|x|\le 1\}}(x)\right)\nu(dx)\right], \tag{1.12}$$

where

$$A \text{ is a symmetric nonnegative-definite } d\times d \text{ matrix}, \tag{1.13}$$

$$\nu \text{ is a measure on } \mathbb{R}^d \text{ with } \nu(\{0\}) = 0 \text{ and } \textstyle\int_{\mathbb{R}^d}(|x|^2\wedge 1)\nu(dx) < \infty, \tag{1.14}$$

$$\gamma \text{ is a vector in } \mathbb{R}^d, \tag{1.15}$$

and A, ν, and γ are uniquely determined by μ. Conversely, if A, ν, and γ satisfy (1.13), (1.14), and (1.15), then there exists a unique $\mu \in ID$ satisfying (1.12).

Definition 1.29 Given $\mu \in ID$, we call A and ν in Theorem 1.28 the *Gaussian covariance matrix* of μ and the *Lévy measure* of μ, respectively. (If $d = 1$, then we often call A the *Gaussian variance* of μ.) We call the triplet (A, ν, γ) the *generating triplet* of μ; sometimes we simply call it the *triplet* of μ. In case ν satisfies $\int_{\mathbb{R}^d} (|x| \wedge 1)\nu(dx) < \infty$, $\gamma^0 \in \mathbb{R}^d$ defined by

$$\gamma^0 = \gamma - \int_{|x|\le 1} x\nu(dx) \tag{1.16}$$

is called the *drift* of μ and (1.12) is written as

$$\widehat{\mu}(z) = \exp\left[-\tfrac{1}{2}\langle z, Az\rangle + i\langle \gamma^0, z\rangle + \int_{\mathbb{R}^d} \left(e^{i\langle z,x\rangle} - 1\right) \nu(dx)\right]. \tag{1.17}$$

A distribution $\mu \in ID$ satisfying (1.17) with $A = 0$, $\gamma^0 = 0$, and $\nu(\mathbb{R}^d) < \infty$ is called *compound Poisson*. A distribution $\mu \in ID$ is *Gaussian* if $\nu = 0$; $\mu \in ID$ is *non-Gaussian* if $\nu \neq 0$; $\mu \in ID$ is *purely non-Gaussian* if $A = 0$ and $\nu \neq 0$. For $d = 1$, $\mu \in ID(\mathbb{R})$ is *Poisson* with parameter $c > 0$ if $A = 0$, $\gamma^0 = 0$, and $\nu = c\delta_1$, which means $\mu(\{n\}) = e^{-c}c^n/(n!)$ for $n = 0, 1, \ldots$.

Notice that a Gaussian distribution on $\mathbb{R}^d$ is possibly *degenerate*, that is, supported on some $(d-1)$-dimensional hyperplane in $\mathbb{R}^d$. Here 0-dimensional hyperplane means a one-point set.

Remark 1.30 Theorem 1.28 shows that, if $\mu \in ID$, then

$$(\log \widehat{\mu})(z) = -\tfrac{1}{2}\langle z, Az\rangle + i\langle \gamma, z\rangle + \int_{\mathbb{R}^d} g(z, x)\nu(dx), \tag{1.18}$$

where

$$g(z, x) = e^{i\langle z,x\rangle} - 1 - i\langle z, x\rangle 1_{\{|x|\le 1\}}(x). \tag{1.19}$$

Indeed, since $g(z, x)$ is continuous in z and satisfies $g(0, x) = 0$ and

$$|g(z, x)| \le \tfrac{1}{2} |z|^2 |x|^2 1_{\{|x|\le 1\}}(x) + 2 \cdot 1_{\{|x|>1\}}(x), \tag{1.20}$$

the dominated convergence theorem shows that the right-hand side of (1.18) is continuous in z and vanishes at $z = 0$.

Definition 1.31 Let $\{X_t : t \ge 0\}$ be a stochastic process on $\mathbb{R}^d$ defined on a probability space $(\Omega, \mathcal{F}, P)$. It is called *Lévy process* if it satisfies the following:

(i) $X_0 = 0$ almost surely.
(ii) For any $0 \le t_0 < \cdots < t_n$, the random variables $X_{t_0}, X_{t_1} - X_{t_0}, \ldots, X_{t_n} - X_{t_{n-1}}$ are independent.
(iii) The distribution of $X_{s+t} - X_s$ does not depend on s.

(iv) For every $t \geq 0$ and $\varepsilon > 0$, $P(|X_s - X_t| \geq \varepsilon) \to 0$ as $s \to t$.
(v) $X_t(\omega)$ is right continuous with left limits in t, almost surely.[5]

We call $\{X_t\}$ an *additive process* if it satisfies (i), (ii), (iv), and (v). Further, we call $\{X_t\}$ an *additive process in law* if it satisfies (i), (ii), and (iv).

We call (ii) and (iii), respectively, the *independent increment property* and the *stationary increment property.* Sometimes we refer to (iv) as *stochastic continuity*.

If $\{X_t\}$ is a Lévy process on $\mathbb{R}^d$, then $\mathcal{L}(X_t)$ is in $ID(\mathbb{R}^d)$ for every $t \geq 0$ and $\mathcal{L}(X_t) = \mu^t$, where $\mu = \mathcal{L}(X_1)$. Further, any joint distribution $\mathcal{L}\left((X_{t_j})_{1\leq j\leq n}\right)$ for $n \in \mathbb{N}$ and $0 \leq t_1 < t_2 < \cdots < t_n$ is infinitely divisible on $\mathbb{R}^{nd}$ and determined by μ.[6] To see this, notice that $X_{t_j} = X_{t_1} + (X_{t_2} - X_{t_1}) + \cdots + (X_{t_j} - X_{t_{j-1}})$, whose distribution is determined by $\{\mu^t : t \geq 0\}$, and that $(X_{t_j})_{1\leq j\leq n}$ is a linear transformation of $(X_{t_j} - X_{t_{j-1}})_{1\leq j\leq n}$ with $t_0 = 0$. Conversely, for every $\mu \in ID(\mathbb{R}^d)$ there exists a Lévy process $\{X_t\}$ such that $\mathcal{L}(X_1) = \mu$. This correspondence between $ID(\mathbb{R}^d)$ and the class of Lévy processes on $\mathbb{R}^d$ is one-to-one and onto if two Lévy processes identical in law are identified.

We use the words *generating triplet*, *Gaussian covariance matrix*, *Lévy measure*, and *drift* of a Lévy process $\{X_t\}$ for those of $\mathcal{L}(X_1)$. A Lévy process $\{X_t\}$ is called *selfdecomposable*, *of class* L_m, *stable*, α-*stable*, *strictly stable*, *strictly* α-*stable*, *Gaussian*, *purely non-Gaussian*, *Poisson*, or *compound Poisson*, respectively, if so is $\mathcal{L}(X_t)$ for all $t \geq 0$. Evidently, this condition is equivalent to saying that "if so is $\mathcal{L}(X_t)$ for some $t > 0$".

For a Lévy process $\{X_t\}$ on $\mathbb{R}^d$, it is known that $X_t(\omega)$ is continuous in t almost surely if and only if it is Gaussian. A Lévy process $\{X_t\}$ on $\mathbb{R}^d$ with generating triplet $(I, 0, 0)$ is called the *Brownian motion* on $\mathbb{R}^d$. Here I is the $d \times d$ identity matrix.

A basic fact on additive processes is the following.

Theorem 1.32

(i) Let $\{X_t\}$ be an additive process on $\mathbb{R}^d$. Then $\mathcal{L}(X_s)$ and $\mathcal{L}(X_t - X_s)$ are in $ID(\mathbb{R}^d)$ for all $0 \leq s \leq t$.
(ii) Let $\{X_t\}$ and $\{X'_t\}$ be additive processes on $\mathbb{R}^d$ such that $X_t \stackrel{d}{=} X'_t$ for every $t \geq 0$. Then $\{X_t\} \stackrel{d}{=} \{X'_t\}$. Namely, for any $n \in \mathbb{N}$ and any choice of $0 \leq t_1 < t_2 < \cdots < t_n$, $(X_{t_j})_{1\leq j\leq n} \stackrel{d}{=} (X'_{t_j})_{1\leq j\leq n}$. Moreover, $\mathcal{L}((X_{t_j})_{1\leq j\leq n}) \in ID(\mathbb{R}^{nd})$.

Proof

(i) We can prove $\mathcal{L}(X_s) \in ID$ as a consequence of Theorem 1.10. To see $\mathcal{L}(X_t - X_s) \in ID$, note that $\{X_{t+s} - X_s : t \geq 0\}$ is an additive process.
(ii) This is shown similarly to the case of Lévy processes above. ■

---

[5] We say that a statement $S(\omega)$ involving ω is true *almost surely* (or *a. s.*) if there is $\Omega_0 \in \mathcal{F}$ with $P[\Omega_0] = 1$ such that $S(\omega)$ is true for all $\omega \in \Omega_0$.

[6] Cone-parameter Lévy processes in Chap. 4 do not have these properties (see Example 4.36).

A remarkable result on path behaviour of an additive process is the Lévy–Itô decomposition, which clarifies the meanings of Lévy measure and Gaussian covariance matrix ([93] Theorems 19.2, 19.3).

The totality of compound Poisson distributions has the following property useful in the theory of infinitely divisible distributions and Lévy processes.

Theorem 1.33 *The totality of compound Poisson distributions on* $\mathbb{R}^d$ *is dense in* $ID(\mathbb{R}^d)$ *in the topology of weak convergence.*

See [93] Corollary 8.8 for a proof.

Now let us turn to selfdecomposable distributions. Let $S = \{\xi \in \mathbb{R}^d \colon |\xi| = 1\}$, the unit sphere in $\mathbb{R}^d$. The following result is proved in [93] Theorem 15.10. Notice that selfdecomposability imposes no restriction on A and γ in the generating triplet (A, ν, γ).

Theorem 1.34 *Let* $\mu \in ID(\mathbb{R}^d)$ *with Lévy measure* ν. *Then,* $\mu \in L_0$ *if and only if*

$$\nu(B) = \int_S \lambda(d\xi) \int_0^\infty 1_B(r\xi) \frac{k_\xi(r)}{r} dr, \quad B \in \mathcal{B}(\mathbb{R}^d), \tag{1.21}$$

where λ *is a finite measure on* S *and* $k_\xi(r)$ *is nonnegative, decreasing in* $r \in (0, \infty)$, *and measurable in* $\xi \in S$.

Remark 1.35 The measure λ and function $k_\xi(r)$ in Theorem 1.34 are not uniquely determined by $\mu \in L_0$. Suppose that $\nu \neq 0$. Then we can choose λ and $k_\xi(r)$ satisfying the additional conditions that $\lambda(S) = 1$, $k_\xi(r)$ is right continuous in r, and

$$\int_0^\infty (r^2 \wedge 1) \frac{k_\xi(r)}{r} dr = c, \tag{1.22}$$

where $c = \int (|x|^2 \wedge 1)\nu(dx) > 0$. If both λ, $k_\xi(r)$ and λ', $k'_\xi(r)$ satisfy (1.21) and these additional conditions, then $\lambda = \lambda'$ and $k_\xi(r) = k'_\xi(r)$ for λ-a. e. ξ. Henceforth we always choose λ and $k_\xi(r)$ satisfying these additional conditions and call λ the *spherical component* of ν and $k_\xi(r)$ the *k-function* of ν (or μ). Define

$$h_\xi(u) = k_\xi(e^u), \quad u \in \mathbb{R}. \tag{1.23}$$

We call $h_\xi(u)$ the *h-function* of ν (or μ). The equality (1.22) is written as

$$\int_{-\infty}^\infty (e^{2u} \wedge 1) h_\xi(u) du = c. \tag{1.24}$$

In the remaining case $\nu = 0$ (that is, μ is Gaussian), we can choose $k_\xi(r) = 0$ in (1.21) and the k-function and the h-function of μ are defined to be 0. Now we have, for any $\mu \in L_0$,

$$\widehat{\mu}(z) = \exp\Bigg[-\tfrac{1}{2}\langle z, Az\rangle + i\langle \gamma, z\rangle + \int_S \lambda(d\xi) \int_0^\infty \left(e^{i\langle z, r\xi\rangle} - 1 - i\langle z, r\xi\rangle 1_{(0,1]}(r)\right) \frac{k_\xi(r)}{r} dr \Bigg]. \tag{1.25}$$

In one dimension ($d = 1$), we have $S = \{+1, -1\}$, $\lambda(\{+1\}) + \lambda(\{-1\}) = 1$, and k-function $k_\xi(r)$ for $\xi = +1, -1$. Hence, letting $k(x) = \lambda(\{+1\})k_{+1}(x)$ for $x > 0$ and $k(x) = \lambda(\{-1\})k_{-1}(-x)$ for $x < 0$, we have, for any $\mu \in L_0(\mathbb{R})$,

$$\widehat{\mu}(z) = \exp\left[-\tfrac{1}{2}Az^2 + i\gamma z + \int_{\mathbb{R}} \left(e^{izx} - 1 - izx 1_{[-1,1]}(x)\right) \frac{k(x)}{|x|} dx \right], \tag{1.26}$$

with $k(x)$ being decreasing and right continuous on $(0, \infty)$ and increasing and left continuous on $(-\infty, 0)$, $k(x) \geq 0$, and $\int_{-\infty}^{\infty}(x^2 \wedge 1)(k(x)/|x|)dx < \infty$. This representation is unique. Sometimes we call $k(x)$ in (1.26) the *k-function* of μ for $d = 1$.

Remark 1.36 The representation of the Lévy measure in Remark 1.35 is a special case of a polar decomposition. In general, if ν is the Lévy measure of $\mu \in ID(\mathbb{R}^d)$ with $\int_{\mathbb{R}^d}(|x|^2 \wedge 1)\nu(dx) = c > 0$, then

$$\nu(B) = \int_S \lambda(d\xi) \int_0^\infty 1_B(r\xi)\nu_\xi(dr), \tag{1.27}$$

where λ is a probability measure on S, ν_ξ is a σ-finite measure on $(0, \infty)$ with $\int_0^\infty (r^2 \wedge 1)\nu_\xi(dr) = c$ for each ξ, and ν_ξ is measurable in ξ (that is, $\nu_\xi(B)$ is measurable in ξ for each $B \in \mathcal{B}((0, \infty))$). Moreover, if both λ, ν_ξ and λ', ν'_ξ give this representation, then $\lambda = \lambda'$ and $\nu_\xi = \nu'_\xi$ for λ-a.e. ξ. Proof is given as an application of the existence (and uniqueness) theorem of conditional distributions.

Definition 1.37 For $\varepsilon > 0$, Δ_ε is the *difference operator*, $\Delta_\varepsilon f(u) = f(u + \varepsilon) - f(u)$. The iteration of Δ_ε n times is denoted by Δ_ε^n. Hence

$$\Delta_\varepsilon^n f(u) = \sum_{j=0}^{n} (-1)^{n-j} \binom{n}{j} f(u + j\varepsilon).$$

Define $\Delta_\varepsilon^0 f = f$. For $n \in \mathbb{Z}_+$ we say that $f(u)$, $u \in \mathbb{R}$, is *monotone of order n* if $(-1)^j \Delta_\varepsilon^j f \geq 0$ for $\varepsilon > 0$ and $j = 0, 1, \ldots, n$. We say that $f(u)$, $u \in \mathbb{R}$, is *completely monotone* if $(-1)^j \Delta_\varepsilon^j f \geq 0$ for $\varepsilon > 0$ and $j \in \mathbb{Z}_+$.

Lemma 1.38

(i) *Let $n \geq 1$. A function $f(u)$ is monotone of order n if and only if, for all $\varepsilon > 0$ and $j = 0, 1, \ldots, n - 1$, $(-1)^j \Delta_\varepsilon^j f$ is decreasing.*

(ii) *Let $n \geq 2$. A function $f(u)$ is monotone of order n if and only if $f \in C^{n-2}$, $(-1)^j f^{(j)} \geq 0$ for $j = 0, 1, \ldots, n-2$, and $(-1)^{n-2} f^{(n-2)}$ is decreasing and convex.*

(iii) *A function $f(u)$ is completely monotone if and only if $f \in C^\infty$ and $(-1)^j f^{(j)} \geq 0$ for $j = 0, 1, \ldots$.*

See Widder [138] pp. 144–151 (1946) for a proof. A consequence of (ii) is that, if $f \in C^n$ and $(-1)^j f^{(j)} \geq 0$ for $j = 0, 1, \ldots, n$, then f is monotone of order n.

Now we will give characterization of the class $L_m\left(\mathbb{R}^d\right)$ in terms of h-functions.

Theorem 1.39

(i) *Let $m \in \{0, 1, \ldots\}$. Then $\mu \in L_m$ if and only if $\mu \in L_0$ and its h-function $h_\xi(u)$ is monotone of order $m+1$ for λ-a. e. ξ.*

(ii) *$\mu \in L_\infty$ if and only if $\mu \in L_0$ and $h_\xi(u)$ is completely monotone for λ-a. e. ξ.*

Proof

(i) The assertion is true for $m = 0$ owing to Theorem 1.34. Let $m \geq 1$. We will show the validity of the assertion for m, assuming that it is valid for $m-1$.

Let $\mu \in L_m$ with Lévy measure ν. If $\nu = 0$, then its h-function 0 is monotone of any order. Let $\nu \neq 0$. We have $\widehat{\mu}(z) = \widehat{\mu}\left(b^{-1}z\right)\widehat{\rho}_b(z)$ for every $b > 1$, and it follows from (1.25) that ρ_b has Lévy measure

$$\nu_b(B) = \int_S \lambda(d\xi) \int_0^\infty 1_B(r\xi) \frac{k_\xi(r) - k_\xi(br)}{r} dr, \quad B \in \mathcal{B}(\mathbb{R}^d).$$

Let $a_b(\xi) = \int_0^\infty (r^2 \wedge 1)\left(k_\xi(r) - k_\xi(br)\right) r^{-1} dr$. We have $0 < a_b(\xi) < c$, where c is as in Remark 1.35. The spherical component λ_b and the h-function $h_{b,\xi}$ of ρ_b are as follows: $\lambda_b(d\xi) = c_b^{-1} a_b(\xi)\lambda(d\xi)$, where c_b is a constant that makes λ_b a probability measure, $k_{b,\xi}(r) = c_b(a_b(\xi))^{-1}\left\{k_\xi(r) - k_\xi(br)\right\}$, and

$$h_{b,\xi}(u) = k_{b,\xi}\left(e^u\right) = c_b(a_b(\xi))^{-1}\left\{h_\xi(u) - h_\xi(u + \log b)\right\}. \tag{1.28}$$

Since $\rho_b \in L_{m-1}$, $h_{b,\xi}$ is monotone of order m for λ-$a.e.$ ξ by the induction hypothesis. Hence

$$(-1)^j\left(\Delta_\varepsilon^j h_\xi(u) - \Delta_\varepsilon^j h_\xi(u + \log b)\right) = (-1)^j c_b^{-1} a_b(\xi)\Delta_\varepsilon^j h_{b,\xi}(u) \geq 0$$

for $j = 0, 1, \ldots, m$. Choosing $b = e^\varepsilon$, we see that $(-1)^j \Delta_\varepsilon^j h_\xi(u) \geq 0$ for $j = 1, 2, \ldots, m+1$, and therefore h_ξ is monotone of order $m+1$ for λ-a.e. ξ.

Conversely, suppose that $\mu \in L_0$ and h_ξ is monotone of order $m+1$ for λ-$a.e.$ ξ. Then, by Lemma 1.38 (i), $(-1)^j \Delta_\varepsilon^j h_\xi(u)$ is decreasing in u for $j = 0, 1, \ldots, m$. Hence it follows from (1.28) that $h_{b,\xi}(u)$ is monotone of order m. Now we have $\rho_b \in L_{m-1}$ from the induction hypothesis. Thus $\mu \in L_m$.

(ii) This is an immediate consequence of (i) and the definition of L_∞. ∎

Next we will consider the class L_∞.

Lemma 1.40

(i) *If $h(u)$ is a completely monotone function on $\mathbb{R}$, then there is a unique measure Π on $[0, \infty)$ such that*

$$h(u) = \int_{[0,\infty)} e^{-\alpha u} \Pi(d\alpha), \quad u \in \mathbb{R}. \tag{1.29}$$

Conversely, if a measure Π on $[0, \infty)$ is such that $\int_{[0,\infty)} e^{-\alpha u} \Pi(d\alpha) < \infty$ for all $u \in \mathbb{R}$, then the function h defined by (1.29) is completely monotone on $\mathbb{R}$.

(ii) *Let $0 < c < \infty$. The function $h(u)$ in (i) satisfies*

$$\int_{-\infty}^{\infty} (e^{2u} \wedge 1) h(u) du = c \tag{1.30}$$

if and only if the corresponding measure Π is concentrated on the open interval $(0, 2)$ and satisfies

$$\int_{(0,2)} \left(\frac{1}{\alpha} + \frac{1}{2 - \alpha} \right) \Pi(d\alpha) = c. \tag{1.31}$$

(iii) *Let $h_\xi(u)$, $\xi \in S$, be completely monotone in $u \in \mathbb{R}$ for each ξ and let Π_ξ be the measure corresponding to h_ξ in (i). Then h_ξ is measurable in ξ if and only if Π_ξ is measurable in ξ.*

Proof

(i) Suppose that $h(u)$ is completely monotone on $\mathbb{R}$. Then by Bernstein's theorem (see Widder [138] (1946) or Feller [24] (1971)) there is, for each u_0, a unique finite measure Π^{u_0} on $[0, \infty)$ such that $h(u_0 + u) = \int_{[0,\infty)} e^{-\alpha u} \Pi^{u_0}(d\alpha)$ for $u \geq 0$. The uniqueness implies that, if $u_0 < u_1$, then $\Pi^{u_1}(d\alpha) = e^{-\alpha(u_1 - u_0)} \Pi^{u_0}(d\alpha)$. Hence Π defined by $\Pi(d\alpha) = e^{\alpha u_0} \Pi^{u_0}(d\alpha)$ does not depend on u_0 and satisfies (1.29). The uniqueness of Π comes from the uniqueness in Bernstein's theorem. Conversely, if Π is a measure satisfying $\int_{[0,\infty)} e^{-\alpha u} \Pi(d\alpha) < \infty$ for all $u \in \mathbb{R}$, then the function h defined by (1.29) is completely monotone on $\mathbb{R}$, since we can change the order of differentiation and integration to obtain

$$(-1)^n (d/du)^n (h(u)) = \int_{[0,2)} e^{-\alpha u} \alpha^n \Pi(d\alpha) \geq 0.$$

(ii) Suppose that (1.30) holds. Then $\Pi(\{0\}) = 0$ since $h(u) \to 0$ as $u \to \infty$. Moreover, we have

$$\int_{(0,\infty)} \Pi(d\alpha) \int_{-\infty}^{0} e^{-u(\alpha - 2)} du = \int_{-\infty}^{0} e^{2u} h(u) du < \infty.$$

Since $\int_{-\infty}^{0} e^{-u(\alpha-2)}du = \infty$ for $\alpha \geq 2$, this implies that Π is concentrated in $(0, 2)$. Now we have (1.31), since

$$c = \int_{-\infty}^{0} e^{2u}h(u)du + \int_{0}^{\infty} h(u)du$$
$$= \int_{(0,2)} \Pi(d\alpha) \int_{-\infty}^{0} e^{u(2-\alpha)}du + \int_{(0,2)} \Pi(d\alpha) \int_{0}^{\infty} e^{-\alpha u}du$$
$$= \int_{(0,2)} \left(\frac{1}{\alpha} + \frac{1}{2-\alpha}\right) \Pi(d\alpha).$$

The "if" part of (ii) is shown similarly.

(iii) The inversion formula for Laplace transforms (see Widder [138] p. 295 (1946) or Feller [24] p. 440 (1971)) says that

$$\int_{0}^{\beta} \Pi_{\xi}(d\alpha) = \lim_{u\to\infty} \sum_{m\leq\beta u} \frac{(-u)^m}{m!} h_{\xi}^{(m)}(u) \tag{1.32}$$

for all $\beta > 0$ satisfying $\Pi_{\xi}(\{\beta\}) = 0$. Hence, for any $\alpha_0 \geq 0$, $\Pi_{\xi}([0, \alpha_0])$ is the limit of the right-hand side of (1.32) as $\beta \downarrow \alpha_0$. This shows the "only if" part of the assertion. The "if" part follows from the equation (1.29) for h_{ξ}. ■

Theorem 1.41

(i) If $\mu \in L_{\infty}$, then

$$\widehat{\mu}(z) = \exp\Bigg[-\tfrac{1}{2}\langle z, Az\rangle + i\langle \gamma, z\rangle$$
$$+ \int_{(0,2)} \Pi(d\alpha) \int_{S} \lambda_{\alpha}(d\xi) \int_{0}^{\infty} \left(e^{i\langle z, r\xi\rangle} - 1 - i\langle z, r\xi\rangle 1_{(0,1]}(r)\right) \frac{dr}{r^{1+\alpha}}\Bigg], \tag{1.33}$$

where A is a nonnegative-definite symmetric $d \times d$ matrix, $\gamma \in \mathbb{R}^d$, Π is a measure on the interval $(0, 2)$ satisfying

$$\int_{(0,2)} \left(\frac{1}{\alpha} + \frac{1}{2-\alpha}\right) \Pi(d\alpha) < \infty, \tag{1.34}$$

λ_{α} is a probability measure on S for each α, and λ_{α} is measurable in α. These A, γ, and Π are uniquely determined by μ; λ_{α} is determined by μ up to α of Π-measure 0.

(ii) Given A, γ, Π, and λ_{α} satisfying the conditions above, we can find $\mu \in L_{\infty}$ satisfying (1.33).

Proof

(i) Suppose that $\mu \in L_\infty$ with Lévy measure ν and h-function h_ξ. Let Π_ξ be the measure in Lemma 1.40 corresponding to h_ξ. Then Π_ξ is measurable in ξ and concentrated in $(0, 2)$, since h_ξ satisfies (1.24). Thus we have

$$h_\xi(u) = \int_{(0,2)} e^{-\alpha u} \Pi_\xi(d\alpha), \tag{1.35}$$

$$\int_{(0,2)} \left(\frac{1}{\alpha} + \frac{1}{2-\alpha}\right) \Pi_\xi(d\alpha) = c. \tag{1.36}$$

Let λ be the spherical component of ν. If $\nu = 0$, then let $\Pi = 0$. If $\nu \neq 0$, then we can find a measure Π on $(0, 2)$ satisfying (1.34) and a probability measure λ_α measurable in α such that

$$\int_{(0,2)} \Pi\,(d\alpha) \int_S \lambda_\alpha(d\xi) f\,(\alpha, \xi) = \int_S \lambda(d\xi) \int_{(0,2)} \Pi_\xi\,(d\alpha)\, f\,(\alpha, \xi) \tag{1.37}$$

for every nonnegative measurable function $f\,(\alpha, \xi)$. In fact, it suffices to apply the existence theorem of conditional distribution to the probability measure given by $c^{-1}\left(\alpha^{-1} + (2-\alpha)^{-1}\right) \lambda(d\xi)\Pi_\xi\,(d\alpha)$ on $(0, 2) \times S$. Now we have

$$\int_{\mathbb{R}^d} f(x)\nu\,(dx) = \int_{(0,2)} \Pi\,(d\alpha) \int_S \lambda_\alpha(d\xi) \int_0^\infty f\,(r\xi)\, r^{-\alpha-1} dr \tag{1.38}$$

for every nonnegative measurable function $f\,(x)$, since (1.21) shows

$$\nu(B) = \int_S \lambda(d\xi) \int_0^\infty 1_B\,(r\xi)\, \frac{h_\xi\,(\log r)}{r} dr, \tag{1.39}$$

from which follows

$$\begin{aligned} \nu(B) &= \int_S \lambda(d\xi) \int_0^\infty 1_B\,(r\xi)\, r^{-1} dr \int_{(0,2)} r^{-\alpha}\, \Pi_\xi\,(d\alpha) \\ &= \int_S \lambda(d\xi) \int_{(0,2)} \Pi_\xi\,(d\alpha) \int_0^\infty 1_B\,(r\xi)\, r^{-\alpha-1} dr \end{aligned}$$

by use of (1.35) and Fubini's theorem. It follows that (1.38) is valid for every complex-valued ν-integrable function f. Letting $f\,(x) = e^{i\langle z,x\rangle} - 1 - i\,\langle z, x\rangle\, 1_{(0,1]}\,(|x|)$ for each z, we get (1.33). Proof of the uniqueness assertion in (i) will be given after the proof of (ii).

(ii) Suppose that A, γ, Π, and λ_α are given and satisfy the conditions in (i). Let c be the value of the integral in (1.34). Then we can find a probability measure λ on S and a measure Π_ξ on $(0, 2)$ measurable in ξ such that (1.36) and (1.37) are satisfied. Define the function h_ξ by (1.35) and let $k_\xi = h_\xi(\log r)$. Define a measure ν by (1.21). Then we can show the equality (1.38) and $\int_S (|x|^2 \wedge 1)\nu(dx) = c$. Now Theorem 1.28 says that there is $\mu \in ID$ with generating triplet (A, γ, ν) and Theorem 1.34 says that this μ is in L_0. Moreover, Theorem 1.39 says that $\mu \in L_\infty$, since $h_\xi(u)$ is completely monotone in $u \in \mathbb{R}$ by Lemma 1.40. This μ satisfies (1.33).

Let us see the uniqueness of A, γ, Π, and λ_α in (i). The expression (1.33) says that μ has generating triplet (A, ν, γ) with ν expressible by (1.38). Construction procedure in the proof of (ii) gives λ and h_ξ, which expresses ν as in (1.39). Hence λ and $h_\xi(\log r)$ are determined by ν, as is stated in Remark 1.35. The h_ξ determines Π_ξ. Now Π and λ_α are determined by λ and Π_ξ. ■

Characterization of α-stable and strictly α-stable distributions on $\mathbb{R}^d$ is as follows ([93] Theorems 14.1, 14.3, and 14.7).

Theorem 1.42

(i) Let $0 < \alpha < 2$. If $\mu \in \mathfrak{S}_\alpha$, then $\mu \in ID$ with generating triplet (A, ν, γ) satisfying $A = 0$ and

$$\nu(B) = c \int_S \lambda(d\xi) \int_0^\infty 1_B(r\xi) r^{-\alpha-1} dr \qquad \text{for every } B \in \mathcal{B}\left(\mathbb{R}^d\right), \tag{1.40}$$

with a probability measure λ on S and a nonnegative constant c. Conversely, for any λ, c, and γ, there is $\mu \in \mathfrak{S}_\alpha$ having generating triplet $(0, \nu, \gamma)$ with ν satisfying (1.40).

(ii) The class $\mathfrak{S}_2$ is the totality of Gaussian distributions.

(iii) Let $0 < \alpha \le 2$ and $\mu \in \mathfrak{S}_\alpha$. Then $\mu \in \mathfrak{S}_\alpha^0$ if and only if one of the following is satisfied.

(a) $0 < \alpha < 1$ and $\gamma^0 = 0$;
(b) $\alpha = 1$ and $\int_S \xi\lambda(d\xi) = 0$;
(c) $1 < \alpha < 2$ and $\gamma + \int_{|x|>1} x\nu(dx) = 0$;[7]
(d) $\alpha = 2$ and $\gamma = 0$.

We see from (1.40) that $\mu \in \mathfrak{S}_\alpha$ with $0 < \alpha < 2$ if and only if $\mu \in L_\infty$ and, in its representation (1.33), $A = 0$ and $\Pi((0, 2) \setminus \{\alpha\}) = 0$.

Theorem 1.43 *The class L_∞ is the smallest class containing $\mathfrak{S}$ and closed under convolution and convergence.*

[7] If $\mu \in \mathfrak{S}_\alpha$ with $1 < \alpha < 2$, then $\int_{|x|>1} |x|\nu(dx) < \infty$, which is shown by (1.40). For any $\mu \in ID$, $\int_{|x|>1} |x|\nu(dx) < \infty$ and $\int_{\mathbb{R}^d} |x|\mu(dx) < \infty$ are equivalent and, in this case, $\int_{\mathbb{R}^d} x\mu(dx) = \gamma + \int_{|x|>1} x\nu(dx)$ ([93] Example 25.12).

Proof It is already shown that the class L_∞ contains $\mathfrak{S}$ and that it is closed under convolution and convergence (Propositions 1.18 and 1.25). Let $\mathfrak{Q}$ be a class containing $\mathfrak{S}$ and closed under convolution and convergence. Let $\mu \in L_\infty$ and look at the representation (1.33) of $\widehat{\mu}$. We will prove that $\mu \in \mathfrak{Q}$. Suppose $\gamma = 0$ and $A = 0$. First, assume that Π is supported by $[\varepsilon, 2-\varepsilon]$ for some positive ε. Let $M\,(d\alpha\, d\xi) = \Pi\,(d\alpha)\,\lambda_\alpha(d\xi)$. It is a finite measure on $[\varepsilon, 2-\varepsilon] \times S$. Choose M_n such that they converge to M and that each M_n is supported by $E_n \times S$ where E_n is a finite set in $[\varepsilon, 2-\varepsilon]$. We have the expression $M_n\,(d\alpha\, d\xi) = \Pi_n\,(d\alpha)\,\lambda_{n,\alpha}(d\xi)$, where Π_n is supported by E_n. Let, for each z,

$$f_z\,(\alpha, \xi) = \int_0^\infty \left(e^{i\langle z, r\xi\rangle} - 1 - i\langle z, r\xi\rangle 1_{(0,1]}(r)\right) \frac{dr}{r^{1+\alpha}}$$

and let μ_n be such that $\widehat{\mu}_n(z) = \exp\left[\int_{(0,2)} \Pi_n(d\alpha) \int_S \lambda_{n,\alpha}(d\xi) f_z\,(\alpha, \xi)\right]$. As μ_n is the convolution of a finite number of stable distributions, it belongs to $\mathfrak{Q}$. Since $f_z\,(\alpha, \xi)$ is continuous in (α, ξ), $\widehat{\mu}_n\,(z)$ converges to $\widehat{\mu}\,(z)$. Hence $\mu \in \mathfrak{Q}$. Next consider a general μ in L_∞. Restrict Π to $\left[n^{-1}, 2-n^{-1}\right]$ and let μ_n be the resulting distribution. Then $\mu_n \in \mathfrak{Q}$, as the Gaussian factor of μ_n is in $\mathfrak{Q}$. Since $\int_{(0,2)} \Pi(d\alpha) \int_S \lambda_\alpha(d\xi)\,|f_z\,(\alpha, \xi)|$ is finite, $\widehat{\mu}_n\,(z)$ tends to $\widehat{\mu}\,(z)$. Hence $\mu \in \mathfrak{Q}$. This concludes the proof. ∎

Remark 1.44 Many properties of selfdecomposable distributions are known. Unimodality of all $\mu \in L_0(\mathbb{R})$ was proved by Yamazato [143] (1978) after several unsuccessful papers. In general, a measure η on $\mathbb{R}$ is called *unimodal* (with mode c) if $\eta = a\delta_c + f(x)dx$, where $0 \le a \le \infty$ and $f(x)$ is increasing on $(-\infty, c)$ and decreasing on (c, ∞). By the result of Yamazato, unimodality of all $\mu \in \mathfrak{S}(\mathbb{R})$ was also established for the first time. Singularity of the density of $\mu \in L_0(\mathbb{R})$ and its degree of smoothness were determined by Sato and Yamazato [108] (1978). Absolute continuity of non-trivial $\mu \in L_0(\mathbb{R})$ is easily observed. Indeed, if $A \neq 0$, then it is because μ has a nondegenerate Gaussian factor; if $A = 0$ and $\nu \neq 0$, then it follows from absolute continuity of ν combined with $\nu(\mathbb{R}) = \infty$. This argument does not work for $\mu \in L_0(\mathbb{R}^d)$ with $d \ge 2$, since ν is not always absolutely continuous even if μ is nondegenerate. However, it is shown by Sato [85] (1982) that any nondegenerate $\mu \in L_0(\mathbb{R}^d)$ is absolutely continuous. See [93] for more accounts.

Example 1.45 Let $\mu_{t,q}$ be Γ*-distribution* on $\mathbb{R}_+ = [0, \infty)$ with shape parameter $t > 0$ and scale parameter $q > 0$, that is,

$$\mu_{t,q}(dx) = \frac{q^t}{\Gamma(t)} x^{t-1} e^{-qx} 1_{(0,\infty)}(x)dx. \tag{1.41}$$

When $t = 1$, $\mu_{1,p}$ is called *exponential distribution* with parameter q. Let us prove that $\mu_{t,q}$ is in $L_0\,(\mathbb{R})$ but not in $L_1\,(\mathbb{R})$. This fact was observed in [8] p. 182. The Laplace transform of $\mu_{t,q}$ is

$$L_{\mu_{t,q}}(u) = \int_0^\infty e^{-ux}\mu_{t,q}(dx) = \frac{q^t}{\Gamma(t)}\int_0^\infty x^{t-1}e^{-(u+q)x}dx$$
$$= \frac{q^t}{(u+q)^t}\int_0^\infty \frac{(u+q)^t}{\Gamma(t)}x^{t-1}e^{-(u+q)x}dx = \left(1+\frac{u}{q}\right)^{-t}, \qquad u \ge 0.$$

On the other hand,

$$\log\left(1+\frac{u}{q}\right) = \int_q^{u+q}\frac{dy}{y} = \int_0^u \frac{dy}{q+y} = \int_0^u dy\int_0^\infty e^{-(q+y)x}dx$$
$$= \int_0^\infty e^{-qx}dx\int_0^u e^{-yx}dy = \int_0^\infty \left(1-e^{-ux}\right)\frac{e^{-qx}}{x}dx.$$

Hence

$$L_{\mu_{t,q}}(u) = \exp\left[t\int_0^\infty \left(e^{-ux}-1\right)\frac{e^{-qx}}{x}dx\right], \qquad u \ge 0.$$

Extending this equality to the left half plane $\{w \in \mathbb{C}\colon \mathrm{Re}(w) \le 0\}$ by analyticity in the interior and continuity to the boundary, we get, for $w = iz$ with $z \in \mathbb{R}$,

$$\widehat{\mu}_{t,q}(z) = \exp\left[t\int_0^\infty \left(e^{izx}-1\right)\frac{e^{-qx}}{x}dx\right]. \tag{1.42}$$

This is a special case of (1.17). Thus $\mu_{t,q} \in ID$ with Lévy measure

$$\nu_{t,q}(dx) = tx^{-1}e^{-qx}1_{(0,\infty)}(x)dx.$$

It follows from Remark 1.35 that $\mu_{t,q} \in L_0$ with k-function $k(x) = te^{-qx}1_{(0,\infty)}(x)$ for fixed t, q. The h-function defined by $h(u) = k(e^u)$, $u \in \mathbb{R}$, equals te^{-qe^u}. Then

$$h'(u) = -tqe^u e^{-qe^u}, \qquad h''(u) = t\left(q^2e^{2u} - qe^u\right)e^{-qe^u}.$$

Note that $h''(u) < 0$ for $u < -\log q$. The h-function is not monotone of order 2 by Lemma 1.38 (ii) because it is not convex on $(-\infty, -\log q)$. Therefore, by Theorem 1.39 (i), $\mu_{t,q} \notin L_1$.

The Lévy process $\{X_t\}$ on $\mathbb{R}$ with $\mathcal{L}(X_1)$ being exponential distribution $\mu_{1,q}$ is called *Γ-Lévy process* or *Γ-process* with parameter q. This process satisfies $\mathcal{L}(X_t) = \mu_{t,q}$, which follows from (1.42). Thus the time parameter and the shape parameter coincide. It is a selfdecomposable process but not a process of class L_1.

Example 1.46 The distribution

$$\mu(dx) = c_0\exp(abx - ce^{ax})dx \tag{1.43}$$

on $\mathbb{R}$ with positive parameters a, b, c and $c_0 = ac^b / \Gamma(b)$ is discussed in Linnik and Ostrovskii [57] p. 52 and p. 361 (1977) (see also [93] E18.19). Here we have made a change of parametrization. It is shown that μ is infinitely divisible with Gaussian variance $A = 0$ and Lévy measure

$$\nu(dx) = |x|^{-1} e^{abx} (1 - e^{ax})^{-1} 1_{(-\infty,0)}(x) dx. \tag{1.44}$$

It follows that $\mu \in L_0$. The support of μ is the whole line but the positive tail of μ is much lighter than the negative tail. (In general the *support* of a measure η on $\mathbb{R}^d$ is defined to be the set of x such that $\eta(G) > 0$ for any open set G containing x.) Another such example is an α-stable distribution with $\alpha \in [1, 2)$ and Lévy measure $|x|^{-\alpha-1} 1_{(-\infty,0)}(x) dx$. These belong to the class of infinitely divisible distributions that some people call *spectrally negative*. The distribution μ has an interesting connection with Γ-process $\{Z_t\}$ with parameter q. Since

$$P[\log Z_t \le u] = \frac{q^t}{\Gamma(t)} \int_{-\infty}^{u} \exp(tx - qe^x) dx,$$

$\log Z_t$ has distribution μ with $a = 1$, $b = t$, and $c = q$. It is proved by Akita and Maejima [1] (2002) that $\mathcal{L}(\log Z_t)$ is in L_1 for $t \ge 1/2$ and in L_2 for $t \ge 1$. The parameter a of μ represents scaling. Indeed, if X is a random variable with distribution μ, then aX has distribution $(c_0/a) \exp(bx - ce^x) dx$. Therefore $\mu \in L_1$ for $b \ge 1/2$ and $\mu \in L_2$ for $b \ge 1$.

We have $\mu \notin L_\infty$ for any choice of a, b, c. Proof is as follows. From (1.43) we see $\int |x|^\alpha \mu(dx) < \infty$ for any $\alpha > 0$, which is equivalent to $\int |x|^\alpha \nu(dx) < \infty$ ([93] Theorem 25.3). But an arbitrary non-Gaussian $\mu \in L_\infty$ with Lévy measure ν satisfies $\int |x|^\alpha \nu(dx) = \infty$ whenever $\alpha \in (0, 2)$ satisfies $\Pi((0, \alpha]) > 0$ for the corresponding measure Π in (1.38).

Example 1.47 The distribution

$$\mu(dx) = (\pi \cosh x)^{-1} dx \tag{1.45}$$

was found by Lévy as the distribution of stochastic area of two-dimensional Brownian motion. He showed that μ is infinitely divisible with generating triplet $(0, \nu, 0)$ where

$$\nu(dx) = (2x \sinh x)^{-1} dx$$

(see [93] Example 15.15). We see from this that μ is selfdecomposable with k-function $k(x) = 2^{-1} |\sinh x|^{-1}$. Let us show that $\mu \in L_1(\mathbb{R})$.

Since $k(x) = k(-x)$, it is enough to consider the function $h(u) = 2^{-1} (\sinh(e^u))^{-1}$ for $u \in \mathbb{R}$. Differentiating twice, we have

$$h'(u) = -2^{-1}e^u \cosh(e^u) \left(\sinh(e^u)\right)^{-2} < 0,$$
$$h''(u) = 2^{-1}e^u \left(\sinh(e^u)\right)^{-3} \left[e^u + 2^{-1}\sinh(2e^u)\left(e^u\coth(e^u) - 1\right)\right].$$

Let $f(x) = x\coth x - 1$. If we can show that $f(x) > 0$ for $x > 0$, then $h''(u) > 0$ and $\mu \in L_1$ by Theorem 1.39 and Lemma 1.38. It suffices to show $x(e^{2x} + 1) > e^{2x} - 1$ for $x > 0$. Let $g(x) = x\left(e^{2x} + 1\right) - e^{2x} + 1$. Then $g'(x) = 1 - e^{2x} + 2xe^{2x}$ and $g''(x) = 4xe^{2x}$. Thus $g(0) = 0$, $g'(0) = 0$, and $g''(x) > 0$ for $x > 0$. Hence $g(x) > 0$ for $x > 0$. Thus $f(x)$ is positive for $x > 0$.

Noting that $\int_{|x|>1} |x|^\alpha \nu(dx) < \infty$ for any $\alpha > 0$, we have $\mu \notin L_\infty$ by the same reason as in Example 1.46. But the highest m for which μ belongs to L_m is not known.

Remark 1.48 Analysis of path behaviours of Lévy processes is thoroughly made by the Lévy–Itô decomposition of sample functions. Some basic facts from this decomposition are the following. Let $\{X_t\}$ be a Lévy process on $\mathbb{R}^d$ with generating triplet (A, ν, γ). Call $s > 0$ a *jump time* of $X_t(\omega)$ if $X_s(\omega) \neq X_{s-}(\omega)$ ($\{X_t\}$ has a *jump* at $s > 0$). Then:

(i) $\int_{|x|\leq 1} \nu(dx) < \infty$ if and only if the number of jump times of $X_t(\omega)$ in any finite time interval is finite a. s.
(ii) $\int_{|x|\leq 1} \nu(dx) = \infty$ if and only if jump times are countable and dense in $(0, \infty)$ a. s.
(iii) $\int_{|x|\leq 1} |x|\nu(dx) < \infty$ if and only if the sum of $|X_s(\omega) - X_{s-}(\omega)|$ over all jump times s in any finite time interval is finite a. s.
(iv) $\int_{|x|\leq 1} |x|\nu(dx) = \infty$ if and only if the sum of $|X_s(\omega) - X_{s-}(\omega)|$ over all jump times s in any finite time interval is infinite a. s.
(v) $A \neq 0$ or $\int_{|x|\leq 1} |x|\nu(dx) = \infty$ if and only if $X_t(\omega)$ has infinite total variation in any finite time interval a. s.
(vi) $A = 0$ and $\int_{|x|\leq 1} |x|\nu(dx) < \infty$ if and only if $X_t(\omega)$ is a linear non-random function plus the sum of all jumps in $(0, t]$ a. s.

Notes

During the pioneering studies of infinitely divisible and stable distributions together with Lévy, additive, and stable processes in 1920s and 1930s, selfdecomposable distributions were, without the name, introduced by Lévy with characterizations similar to Theorem 1.26 (i) and Theorem 1.34. It was in answer to a question raised by Khintchine, who named the distributions as of class L. Expositions of the results were given in the books by Lévy [54] (1937), Khintchine [46] (1938), Gnedenko and Kolmogorov [26] (1968), and Loève [58] (1977, 1978).

The classes L_m, $m = 1, 2, \ldots$, and L_∞ were introduced by Urbanik [129] (1972b), [130] (1973), as a decreasing sequence of classes which possess stronger similarity in some sense to the class of stable distributions as m increases. Then Kumar and Schreiber [49] (1978), [50] (1979) and Thu [122] (1979) followed. Theorems 1.26, 1.39, and 1.41 are reformulation and extension of Urbanik's theory by Sato [84] (1980).

The definition (1.23) of h-function follows Sato [101] (2010); it is different from that of [84] and the former edition of this book. The meaning of "monotone of order m" is also different from that of [84] (1980).

Selfdecomposable and stable distributions and processes are extensively studied; many results are known, see Bertoin [12] (1996) and Sato [93, 95] (1999, 2001b). If $\mu \in \mathfrak{P}(\mathbb{R}^d)$ satisfies (1.3) for some $b \in (1, \infty)$ and some $\rho_b \in ID(\mathbb{R}^d)$, then $\mu \in ID$ and it is called semi-selfdecomposable with b as a span. If $\mu \in ID(\mathbb{R}^d)$ satisfies (1.7) for some $\alpha \in (0, 2)$ and some $b \in (1, \infty)$, then μ is called α-semi-stable with b as a span. If a Lévy process $\{X_t\}$ is such that $\mathcal{L}(X_1)$ is semi-selfdecomposable (resp. α-semi-stable) with b as a span, then $\{X_t\}$ is called a semi-selfdecomposable process (resp. an α-semi-stable) with b as a span. Results are extended to these classes in weaker forms. Some on semi-selfdecomposable distributions are found in Maejima and Sato [60] (2003) and Maejima and Ueda [64] (2009).

Unimodality and non-unimodality can have time evolution. Among Lévy processes on $\mathbb{R}$ some have time evolution from unimodal to non-unimodal and some have time evolution from non-unimodal to unimodal. Furthermore, for any $\alpha \in (0, 1)$, there is an α-semi-stable subordinator (see Definition 4.1 for subordinator) with b as a span such that, for some $t_1 > 0$ and $t_2 > 0$, $\mathcal{L}(X_t)$ is unimodal for $t = b^n t_1$ for all $n \in \mathbb{Z}$ and non-unimodal for $t = b^n t_2$ for all $n \in \mathbb{Z}$. Notice that both $b^n t_1$ and $b^n t_2$ accumulate at 0 and ∞. Time evolution in modality of several Lévy processes on $\mathbb{R}$ with explicit Lévy measures is investigated; see Sato [91] (1997) and Watanabe [134, 136] (1999, 2001).

A measure η on $\mathbb{R}^d$ is called *discrete* if there is a countable set $C \in \mathcal{B}(\mathbb{R}^d)$ such that $\eta(\mathbb{R}^d \setminus C) = 0$; *continuous* if $\eta(\{x\}) = 0$ for all $x \in \mathbb{R}^d$; *absolutely continuous* if $\eta(B) = 0$ for all $B \in \mathcal{B}(\mathbb{R}^d)$ having Lebesgue measure 0; *singular* if there is $B \in \mathcal{B}(\mathbb{R}^d)$ with Lebesgue measure 0 such that $\eta(\mathbb{R}^d \setminus B) = 0$; *continuous singular* if it is continuous and singular. Concerning these continuity properties, there are two basic theorems. Let $\mu \in ID(\mathbb{R}^d)$ with generating triplet (A, ν, γ). The continuity theorem says that μ is continuous if and only if $A \neq 0$ or $\nu(\mathbb{R}^d) = \infty$ (see [93] Theorem 27.4). The Hartman–Wintner theorem in [32] (1942) says that, if $A = 0$, ν is discrete, and $\nu(\mathbb{R}^d) = \infty$, then μ is either absolutely continuous or continuous singular (see [93] Theorem 27.16).

Orey [69] (1968) showed the following. If $\{X_t\}$ is a Lévy process on $\mathbb{R}$ with generating triplet (A, ν, γ) such that $A = 0$ and $\nu = \sum_{n=1}^{\infty} a_n^{-\alpha} \delta_{a_n}$ with $a_n = 2^{-c^n}$ where $0 < \alpha < 2$, $c \in \mathbb{N}$, and $c > 2/(2 - \alpha)$, then $\mathcal{L}(X_t)$ is continuous singular for every $t > 0$. The proof uses the Hartman–Wintner theorem and the Riemann–Lebesgue theorem (that is, if $\mu \in \mathfrak{P}(\mathbb{R}^d)$ is absolutely continuous, then $|\widehat{\mu}(z)| \to 0$ as $|z| \to \infty$). This seems to be the simplest example of such a Lévy process and a continuous singular distribution in $ID(\mathbb{R})$; both finite and infinite cases of

$\int_{(0,1)} x\nu(dx)$ are possible, since $\int_{(0,1)} x^{\alpha}\nu(dx) = \infty$ and $\int_{(0,1)} x^{\alpha'}\nu(dx) < \infty$ for $\alpha' > \alpha$ (see [93] E 29.12).

There exists a Lévy process on $\mathbb{R}$ such that, for some $t_0 \in (0, \infty)$, $\mathcal{L}(X_t)$ is continuous singular for all $t \in (0, t_0)$ and absolutely continuous for all $t \in (t_0, \infty)$. This is a remarkable time evolution of a qualitative distributional property of a Lévy process; it is a sort of phase transition. Two methods of construction of such a Lévy process are known; see Tucker [126] (1965), Sato [90] (1994), and Watanabe [135, 136] (2000, 2001); the Lévy process constructed by either method has discrete Lévy measure ν satisfying $\int_{|x|<1} |x|\nu(dx) < \infty$. On the other hand, the continuity theorem implies that there is no time evolution on continuity for any Lévy process $\{X_t\}$; that is, if $\mathcal{L}(X_t)$ is continuous for some $t \in (0, \infty)$, then it is continuous for all $t \in (0, \infty)$.

Chapter 2
Classes L_m and Ornstein–Uhlenbeck Type Processes

Any non-zero Lévy process does not have limit distribution as $t \to \infty$. But the Ornstein–Uhlenbeck process constructed from Brownian motion has a limit distribution as $t \to \infty$, which is Gaussian. Processes of Ornstein–Uhlenbeck type are analogues of the Ornstein–Uhlenbeck process with the role of Brownian motion played by general Lévy processes. In this chapter we shall construct them, give the condition under which they have limit distributions, and study the connection with classes L_m.

In Sect. 2.1, for a Lévy process $\{Z_t\}$ on $\mathbb{R}^d$, we define stochastic integrals based on $\{Z_t\}$ of deterministic bounded measurable functions on a bounded time interval. They are given as limits in probability of stochastic integrals of step functions. Their distributions are described and a Fubini type theorem is proved.

In Sect. 2.2 we define an Ornstein–Uhlenbeck type process $\{X_t\}$ on $\mathbb{R}^d$ as a temporally homogeneous Markov process with a transition function of a special form described by $\rho \in ID$ and $c > 0$. Intuitively speaking, $\{X_t\}$ is a Lévy process $\{Z_t\}$ with $\mathcal{L}(Z_1) = \rho$ to which a force is exerted toward the origin with magnitude c times the distance to the origin. Such a process is constructed as a unique solution of some stochastic integral equation involving $\{Z_t\}$ and c. The Ornstein–Uhlenbeck type process has a limit distribution under an integrability condition on the Lévy measure ν_ρ of ρ. Specifically, if ν_ρ satisfies

$$\int_{|x|>2} \log |x| \, \nu_\rho (dx) < \infty, \tag{2.1}$$

then, as $t \to \infty$, $\mathcal{L}(X_t)$ converges to a selfdecomposable distribution μ on $\mathbb{R}^d$. Conversely, every selfdecomposable distribution μ appears as the limit distribution of some Ornstein–Uhlenbeck type process. If condition (2.1) does not hold, $\mathcal{L}(X_t)$ does not tend to any distribution as t tends to ∞.

A. Rocha-Arteaga, K. Sato, *Topics in Infinitely Divisible Distributions and Lévy Processes*, SpringerBriefs in Probability and Mathematical Statistics,
https://doi.org/10.1007/978-3-030-22700-5_2

In Sect. 2.3 it is shown that there is a one-to-one and onto correspondence between the Lévy processes of class L_{m-1} satisfying (2.1) on the one hand, and the distributions of class L_m which appear as the limit distributions of the Ornstein–Uhlenbeck type processes on the other hand. This correspondence preserves α-stability.

2.1 Stochastic Integrals Based on Lévy Processes

In this section let $\{Z_t : t \geq 0\}$ be a Lévy process on $\mathbb{R}^d$ with $\mathcal{L}(Z_1) = \rho$ and

$$Ee^{i\langle z, Z_t\rangle} = \left(\widehat{\rho^t}\right)(z) = e^{t\psi_\rho(z)}, \tag{2.2}$$

where $\psi_\rho(z)$ is the distinguished logarithm of $\widehat{\rho}$. Further, let $(A_\rho, \nu_\rho, \gamma_\rho)$ be the generating triplet of $\{Z_t\}$. Sometimes $\{Z_t\}$ is denoted by $\{Z_t^{(\rho)}\}$, which expresses that $\mathcal{L}(Z_1) = \rho$. Let us define the stochastic integral of a bounded measurable function defined on a bounded closed interval in $\mathbb{R}_+$ based on $\{Z_t\}$. We will describe the characteristic function of the integral.

Definition 2.1 Let $0 \leq t_0 < t_1 < \infty$. A function $f(s)$ on $[t_0, t_1]$ is called a *step function* if there are a finite number of points $t_0 = s_0 < s_1 < \cdots < s_n = t_1$ such that

$$f(s) = \sum_{j=1}^{n} a_j 1_{[s_{j-1}, s_j)}(s) \tag{2.3}$$

with some $a_1, \ldots, a_n \in \mathbb{R}$. When $f(s)$ is a step function of this form, define

$$\int_{t_0}^{t_1} f(s) dZ_s = \sum_{j=1}^{n} a_j (Z_{s_j} - Z_{s_{j-1}}). \tag{2.4}$$

It follows that $\int_{t_0}^{t_1} f(s) dZ_s$ has infinitely divisible distribution and

$$E\exp\left[i\left\langle z, \int_{t_0}^{t_1} f(s)dZ_s\right\rangle\right] = \prod_{j=1}^{n} E\exp\left[i\left\langle a_j z, Z_{s_j} - Z_{s_{j-1}}\right\rangle\right]$$

$$= \prod_{j=1}^{n} \exp\left[(s_j - s_{j-1})\,\psi_\rho(a_j z)\right] = \exp\left[\sum_{j=1}^{n} (s_j - s_{j-1})\,\psi_\rho(a_j z)\right]$$

$$= \exp\int_{t_0}^{t_1} \psi_\rho(f(s)\,z)\,ds. \tag{2.5}$$

Proposition 2.2 *Let $f(s)$ be a real-valued bounded measurable function on $[t_0, t_1]$ such that there are uniformly bounded step functions $f_n(s)$, $n = 1, 2, \ldots$, on $[t_0, t_1]$ satisfying $f_n \to f$ almost everywhere. Then $\int_{t_0}^{t_1} f_n(s)dZ_s$ converges to an $\mathbb{R}^d$-valued random variable X in probability. The limit X does not depend on the choice of f_n almost surely. The law of X is infinitely divisible and represented as*

$$Ee^{i\langle z,X\rangle} = \exp \int_{t_0}^{t_1} \psi_\rho(f(s)z)ds. \tag{2.6}$$

Using this result, we give

Definition 2.3 The $\mathbb{R}^d$-valued random variable X in Proposition 2.2 is called *stochastic integral* of f on $[t_0, t_1]$ based on the Lévy process $\{Z_s\} = \{Z_s^{(\rho)}\}$ and denoted by

$$X = \int_{t_0}^{t_1} f(s)dZ_s = \int_{t_0}^{t_1} f(s)dZ_s^{(\rho)}. \tag{2.7}$$

Proof of Proposition 2.2 Since ψ_ρ is continuous and $\psi_\rho(0) = 0$, $\psi_\rho((f_n(s) - f_m(s))z)$ tends to 0 for almost every s, as $n, m \to \infty$. Then

$$\int_{t_0}^{t_1} \psi_\rho\left((f_n(s) - f_m(s))\, z\right) ds \to 0$$

as $n, m \to \infty$. Hence by (2.5)

$$\int_{t_0}^{t_1} f_n(s)dZ_s - \int_{t_0}^{t_1} f_m(s)dZ_s = \int_{t_0}^{t_1} (f_n(s) - f_m(s))\, dZ_s \to 0$$

in probability. It follows that there exists a random variable X which is the limit in probability of $\int_{t_0}^{t_1} f_n(s)dZ_s$. The law of X is infinitely divisible, since the laws of $\int_{t_0}^{t_1} f_n(s)dZ_s$ are. Moreover,

$$\int_{t_0}^{t_1} \psi_\rho\left(f_n(s)\, z\right) ds \to \int_{t_0}^{t_1} \psi_\rho\left(f(s)\, z\right) ds$$

by Lebesgue's bounded convergence theorem. Then, by (2.5)

$$E \exp\left[i\left\langle z, \int_{t_0}^{t_1} f_n(s)dZ_s\right\rangle\right] \to \exp \int_{t_0}^{t_1} \psi_\rho\left(f(s)\, z\right) ds.$$

Hence we have (2.6). To see that the limit X does not depend on approximating sequences of f, let $f_n(s) \to f(s)$ and $g_n(s) \to f(s)$ a.e. both boundedly. Then

$$Ee^{i\left\langle z, \int_{t_0}^{t_1} (f_n - g_n) dZ_s \right\rangle} = \exp \int_{t_0}^{t_1} \psi_\rho ((f_n(s) - g_n(s)) z) \, ds \to 1$$

as $n \to \infty$, showing that $\int_{t_0}^{t_1} f_n dZ_s - \int_{t_0}^{t_1} g_n dZ_s \to 0$ in probability. ■

Proposition 2.4 *If $f(s)$ is a real-valued bounded measurable function on $[t_0, t_1]$, then $\int_{t_0}^{t_1} f(s) dZ_s$ is definable in the sense of (2.7) and we have (2.6).*

Proof By Proposition 2.2, it is enough to show the existence of uniformly bounded step functions $f_n(s)$ such that $f_n(s) \to f(s)$ a. e. Let $|f(s)| \le C$. By Lusin's theorem ([31, p. 243]), for each n, there is a closed set $F_n \subset [t_0, t_1]$ such that $[t_0, t_1] \setminus F_n$ has Lebesgue measure less than 2^{-n} and the restriction of f to F_n is continuous. Then, by Urysohn's theorem in general topology, there is a continuous function g_n on $[t_0, t_1]$ with $|g_n(s)| \le C$ such that $g_n = f$ on F_n. We can choose step functions f_n on $[t_0, t_1]$ such that $|f_n(s) - g_n(s)| < 2^{-n}$ and $|f_n(s)| \le C$. Let $G = \bigcap_{k=1}^{\infty} \bigcup_{n=k}^{\infty} ([t_0, t_1] \setminus F_n)$. Then G has Lebesgue measure 0. If $s \notin G$, then $s \in \bigcup_{k=1}^{\infty} \bigcap_{n=k}^{\infty} F_n$ and $f_n(s) \to f(s)$, since

$$|f_n(s) - f(s)| = |f_n(s) - g_n(s)| < 2^{-n}$$

for all large n. ■

Proposition 2.5 *Let $f(s)$ be a locally bounded, measurable function on $[0, \infty)$. Then there is an additive process $\{X_t : t \ge 0\}$ on $\mathbb{R}^d$ such that, for every $t > 0$,*

$$P\left[X_t = \int_0^t f(s) dZ_s \right] = 1. \tag{2.8}$$

Proof Let $Y_0 = 0$ and $Y_t = \int_0^t f(s) dZ_s$ for $t > 0$. If $0 \le t_0 < t_1 < t_2$, then

$$\int_{t_0}^{t_1} f(s) dZ_s + \int_{t_1}^{t_2} f(s) dZ_s = \int_{t_0}^{t_2} f(s) dZ_s \qquad \text{a. s.},$$

as is proved from the case of step functions. This combined with the independent increment property of $\{Z_t\}$ proves that $\{Y_t\}$ has independent increments. If $t_n \downarrow t$, then

$$Ee^{i\langle z, Y_{t_n} - Y_t \rangle} = Ee^{i\langle z, \int_t^{t_n} f(s) dZ_s \rangle} = \exp \int_t^{t_n} \psi_\rho (f(s) z) \, ds \to 1$$

and, similarly, if $t > 0$ and $t_n \uparrow t$, then $Ee^{i\langle z, Y_t - Y_{t_n} \rangle} \to 1$. Hence $\{Y_t\}$ is stochastically continuous. This shows that $\{Y_t\}$ is an additive process in law in Definition 1.31. Now, $\{Y_t\}$ has a modification $\{X_t\}$ which is an additive process, by Theorem 11.5 of [93]. ■

Remark 2.6 Henceforth, $\int_0^t f(s)dZ_s$ is understood to be the modification X_t in Proposition 2.5. For $0 < t_0 \le t$, $\int_{t_0}^t f(s)dZ_s$ is understood to be $X_t - X_{t_0}$.

We need a Fubini type theorem for stochastic integrals and ordinary integrals in order to prove the existence of Ornstein–Uhlenbeck type processes. We establish the following proposition.

Proposition 2.7 *Let $f(s)$ and $g(s)$ be bounded measurable functions on $[t_0, t_1]$. Then*

$$\int_{t_0}^{t_1} g(s)ds \int_{t_0}^{s} f(u)dZ_u = \int_{t_0}^{t_1} f(u)dZ_u \int_{u}^{t_1} g(s)ds \quad a.s., \tag{2.9}$$

which is another writing of

$$\int_{t_0}^{t_1} g(s)\left(\int_{t_0}^{s} f(u)dZ_u\right)ds = \int_{t_0}^{t_1} f(u)\left(\int_{u}^{t_1} g(s)ds\right)dZ_u \quad a.s.$$

Proof Let $X = X(f,g)$ and $Y = Y(f,g)$ denote the left- and the right-hand side of (2.9), respectively. Existence of Y follows from Proposition 2.4. For any additive process $\{X_t\}$, $X_t(\omega)$ is bounded and measurable in $t \in [t_0, t_1]$ for almost every ω by virtue of (v) in Definition 1.31. Hence X is definable by Proposition 2.5.

Step 1. We show that

$$Ee^{i\langle z,X\rangle} = Ee^{i\langle z,Y\rangle} = \exp\int_{t_0}^{t_1} \psi_\rho\left(f(u)\int_{u}^{t_1} g(s)ds\, z\right)du. \tag{2.10}$$

The second equality is evident from (2.6). Let us calculate $Ee^{i\langle z,X\rangle}$. Let $t_{n,k} = t_0 + k2^{-n}(t_1 - t_0)$ for $n = 1, 2, \ldots$ and $k = 0, 1, \ldots, 2^n$. For $s \in [t_0, t_1)$, define $\lambda_n(s) = t_{n,k}$ if $t_{n,k-1} \le s < t_{n,k}$. Let

$$X_n = \int_{t_0}^{t_1} g(s)ds \int_{t_0}^{\lambda_n(s)} f(u)dZ_u.$$

Since $\int_{t_0}^{s} f(u)dZ_u$ is right continuous and locally bounded in s a.s., X_n tends to X a.s. as $n \to \infty$. Hence $Ee^{i\langle z,X_n\rangle} \to Ee^{i\langle z,X\rangle}$. We have

$$X_n = \sum_{k=1}^{2^n} c_k \int_{t_0}^{t_{n,k}} f(u)dZ_u = \int_{t_0}^{t_1} \sum_{k=1}^{2^n} c_k 1_{[t_0,t_{n,k})}(u) f(u)dZ_u \quad \text{a.s.}$$

with $c_k = \int_{t_{n,k-1}}^{t_{n,k}} g(s)ds$. Thus, by (2.6),

$$Ee^{i\langle z,X_n\rangle} = \exp\int_{t_0}^{t_1} \psi_\rho\left(\sum_{k=1}^{2^n} c_k 1_{[t_0,t_{n,k})}(u) f(u)\, z\right) du$$

$$= \exp\int_{t_0}^{t_1} \psi_\rho\left(\sum_{k=1}^{2^n} 1_{[t_0,t_{n,k})}(u)\int_{t_{n,k-1}}^{t_{n,k}} g(s)dsf(u)\, z\right) du$$

$$= \exp\int_{t_0}^{t_1} \psi_\rho\left(\int_{\lambda_n(u)-2^{-n}(t_1-t_0)}^{t_1} g(s)dsf(u)\, z\right) du,$$

which tends to the rightmost member of (2.10) as $n \to \infty$.

Step 2. Let us show that $X = Y$ a.s., assuming that f and g are step functions. Without loss of generality, we can assume that

$$f(s) = \sum_{j=1}^{N} a_j 1_{[s_{j-1},s_j)}(s), \qquad g(s) = \sum_{j=1}^{N} b_j 1_{[s_{j-1},s_j)}(s)$$

with $t_0 = s_0 < s_1 < \cdots < s_N = t_1$. First we prepare the identity

$$\int_{t_0}^{t_1} sdZ_s = t_1 Z_{t_1} - t_0 Z_{t_0} - \int_{t_0}^{t_1} Z_s ds \quad \text{a. s.} \tag{2.11}$$

Define $t_{n,k}$ and $\lambda_n(s)$ as in Step 1. Since $\lambda_n(s)$, $n = 1, 2, \ldots$, are step functions and $\lambda_n(s) \to s$, we have $\int_{t_0}^{t_1} \lambda_n(s)dZ_s \to \int_{t_0}^{t_1} sdZ_s$ in probability. Notice that

$$\int_{t_0}^{t_1} \lambda_n(s)dZ_s = \sum_{k=1}^{2^n} t_{n,k}\left(Z_{t_{n,k}} - Z_{t_{n,k-1}}\right)$$

$$= \sum_{k=1}^{2^n} t_{n,k} Z_{t_{n,k}} - \sum_{k=0}^{2^n-1} \left(t_{n,k} + 2^{-n}(t_1 - t_0)\right) Z_{t_{n,k}}$$

$$= t_1 Z_{t_1} - t_0 Z_{t_0} - \sum_{k=0}^{2^n-1} 2^{-n}(t_1 - t_0) Z_{t_{n,k}}$$

$$= t_1 Z_{t_1} - t_0 Z_{t_0} - \int_{t_0}^{t_1} Z_{\lambda_n(s)} ds + 2^{-n}(t_1 - t_0)\left(Z_{t_1} - Z_{t_0}\right)$$

$$\to t_1 Z_{t_1} - t_0 Z_{t_0} - \int_{t_0}^{t_1} Z_s ds \quad \text{a. s.}$$

as $n \to \infty$, since $Z_{\lambda_n(s)} \to Z_s$ boundedly on $[t_0, t_1)$ a. s. This proves (2.11). Now

$$X = \sum_{k=1}^{N} b_k \int_{s_{k-1}}^{s_k} ds \int_{t_0}^{s} f(u) dZ_u$$

$$= \sum_{k=1}^{N} b_k \int_{s_{k-1}}^{s_k} \sum_{j=1}^{N} a_j \left(Z_{s\wedge s_j} - Z_{s\wedge s_{j-1}}\right) ds = I_1, \text{ say.}$$

Since

$$\int_{s_{k-1}}^{s_k} \left(Z_{s\wedge s_j} - Z_{s\wedge s_{j-1}}\right) ds = \begin{cases} 0 & \text{for } k \le j-1 \\ \int_{s_{k-1}}^{s_k} \left(Z_s - Z_{s_{k-1}}\right) ds & \text{for } k = j \\ \left(Z_{s_j} - Z_{s_{j-1}}\right)(s_k - s_{k-1}) & \text{for } k \ge j+1, \end{cases}$$

we have

$$I_1 = \sum_{j=1}^{N} a_j \left(b_j \int_{s_{j-1}}^{s_j} \left(Z_s - Z_{s_{j-1}}\right) ds + \sum_{k=j+1}^{N} b_k \left(Z_{s_j} - Z_{s_{j-1}}\right)(s_k - s_{k-1}) \right)$$

$$= \sum_{j=1}^{N} a_j \left(-b_j \int_{s_{j-1}}^{s_j} s dZ_s + \left(Z_{s_j} - Z_{s_{j-1}}\right) \left(b_j s_j + \sum_{k=j+1}^{N} b_k (s_k - s_{k-1}) \right) \right).$$

Here we used (2.11). On the other hand,

$$Y = \sum_{j=1}^{N} a_j \int_{s_{j-1}}^{s_j} dZ_u \int_{u}^{t_1} g(s) ds$$

$$= \sum_{j=1}^{N} a_j \int_{s_{j-1}}^{s_j} \left(b_j (s_j - u) + \sum_{k=j+1}^{N} b_k (s_k - s_{k-1}) \right) dZ_u$$

$$= \sum_{j=1}^{N} a_j \left(b_j \int_{s_{j-1}}^{s_j} (s_j - u) dZ_u + \left(Z_{s_j} - Z_{s_{j-1}}\right) \sum_{k=j+1}^{N} b_k (s_k - s_{k-1}) \right).$$

Therefore $X = Y$ a. s. whenever f and g are step functions.

Step 3. We show $X = Y$ a. s. when f is bounded measurable and g is a step function. By Proposition 2.4 there are uniformly bounded step functions f_n such that $f_n \to f$ a.e. on $[t_0, t_1]$. Let $X_n = X(f_n, g)$ and $Y_n = Y(f_n, g)$. We have $X_n = Y_n$ a. s. by Step 2. Since $X - X_n = X(f - f_n, g)$ and $Y - Y_n = Y(f - f_n, g)$, Step 1 gives

$$Ee^{i\langle z,X-X_n\rangle} = Ee^{i\langle z,Y-Y_n\rangle} = \exp\int_{t_0}^{t_1} \psi_\rho\left((f(u)-f_n(u))\int_u^{t_1} g(s)dsz\right)du,$$

which tends to 1 as $n \to \infty$. It follows that $X_n \to X$ and $Y_n \to Y$ in probability. Therefore $X = Y$ a. s.

Step 4. Now let us show $X = Y$ a.s. when f and g are bounded measurable functions. Choose uniformly bounded step functions g_n such that $g_n \to g$ a. e. on $[t_0, t_1]$. Let $\widetilde{X}_n = X(f, g_n)$ and $\widetilde{Y}_n = Y(f, g_n)$. Then $\widetilde{X}_n = \widetilde{Y}_n$ by Step 3 and, by the same method as in Step 3, we can show that $\widetilde{X}_n \to X$ and $\widetilde{Y}_n \to Y$ in probability. Hence (2.9) is proved. ■

Example 2.8 Let $f(s)$ be of class C^1 on $[t_0, t_1]$. As an example of the use of Proposition 2.7, let us show the integration-by-parts formula

$$\int_{t_0}^{t_1} f(u)dZ_u = f(t_1)Z_{t_1} - f(t_0)Z_{t_0} - \int_{t_0}^{t_1} Z_s f'(s)ds \quad \text{a. s.} \tag{2.12}$$

Indeed, notice that

$$\begin{aligned}
\int_{t_0}^{t_1} f(u)dZ_u &= -\int_{t_0}^{t_1} dZ_u \int_u^{t_1} f'(s)ds + f(t_1)\int_{t_0}^{t_1} dZ_u \\
&= -\int_{t_0}^{t_1} f'(s)ds \int_{t_0}^{s} dZ_u + f(t_1)\int_{t_0}^{t_1} dZ_u \qquad \text{(by Proposition 2.7)} \\
&= -\int_{t_0}^{t_1} f'(s)\left(Z_s - Z_{t_0}\right)ds + f(t_1)\left(Z_{t_1} - Z_{t_0}\right) \\
&= -\int_{t_0}^{t_1} f'(s)Z_s ds + (f(t_1) - f(t_0))Z_{t_0} + f(t_1)\left(Z_{t_1} - Z_{t_0}\right),
\end{aligned}$$

which is the right-hand side of (2.12).

2.2 Ornstein–Uhlenbeck Type Processes and Limit Distributions

Let $\{Z_t : t \geq 0\} = \{Z_t^{(\rho)} : t \geq 0\}$ be a Lévy process on $\mathbb{R}^d$ with $\mathcal{L}(Z_1) = \rho$. Let $(A_\rho, \nu_\rho, \gamma_\rho)$ be the generating triplet of ρ and let $\psi_\rho(z)$ be the distinguished logarithm of $\widehat{\rho}$. Let J be a random variable on $\mathbb{R}^d$. In the following we always assume that J and $\{Z_t\}$ are independent. Given $c \in \mathbb{R}$, consider the equation

$$X_t = J + Z_t - c\int_0^t X_s ds, \quad t \geq 0. \tag{2.13}$$

A stochastic process $\{X_t : t \geq 0\}$ is said to be a *solution* of (2.13) if $X_t(\omega)$ is right continuous with left limits in t and satisfies (2.13) a. s.

Proposition 2.9 *Equation* (2.13) *has an almost surely unique solution* $\{X_t\}$ *and, almost surely,*

$$X_t = e^{-ct} J + e^{-ct} \int_0^t e^{cs} dZ_s, \quad t \geq 0. \tag{2.14}$$

Proof Define X_t by (2.14). Then

$$\begin{aligned} c \int_0^t X_s ds &= Jc \int_0^t e^{-cs} ds + c \int_0^t e^{-cs} ds \int_0^s e^{cu} dZ_u \\ &= J\left(1 - e^{-ct}\right) + c \int_0^t e^{cu} dZ_u \int_u^t e^{-cs} ds \\ &= J\left(1 - e^{-ct}\right) + \int_0^t \left(1 - e^{-c(t-u)}\right) dZ_u \\ &= J - X_t + Z_t, \end{aligned}$$

where we have applied Proposition 2.7. Therefore (2.13) holds.

Let us prove the uniqueness of the solution of (2.13). Suppose that $X_t^1(\omega)$ and $X_t^2(\omega)$ satisfy (2.13). For a fixed ω define a bounded function $f(t)$ on $[t_0, t_1]$ by $f(t) = X_t^1(\omega) - X_t^2(\omega)$. Then we have $f(t) = -c \int_0^t f(s)\, ds$ and

$$\begin{aligned} f(t) &= -c \int_0^t \left(-c \int_0^s f(u)\, du \right) ds = (-c)^2 \int_0^t f(u) du \int_u^t ds \\ &= (-c)^2 \int_0^t (t-s) f(s)\, ds. \end{aligned}$$

By induction we get

$$f(t) = \frac{(-c)^n}{(n-1)!} \int_0^t (t-s)^{n-1} f(s)\, ds, \qquad \text{for } n = 1, 2, 3, \ldots$$

Since $\sum_{n=1}^{\infty} \left(|-c|^n / (n-1)!\right) (t-s)^{n-1}$ is finite, $\left(|-c|^n / (n-1)!\right) (t-s)^{n-1}$ tends to 0 uniformly in $s \in [0, t]$ as $n \to \infty$. Hence $f(t) = 0$. This concludes the proof. ■

Let us define a Markov process, using conditional probability.

Definition 2.10 Let T be an interval in $\mathbb{R}$. Let $\{X_t : t \in T\}$ be a stochastic process on $\mathbb{R}^d$ defined on a probability space $(\Omega, \mathcal{F}, P)$. Let, for $t \in T$, $\mathcal{F}_t$ be the σ-algebra generated by $\{X_s : s \in T \cap (-\infty, t]\}$ (that is, $\mathcal{F}_t$ is the smallest sub-σ-algebra of $\mathcal{F}$ that contains $\{\omega \in \Omega : X_s(\omega) \in B\}$ for all $s \in T \cap (-\infty, t]$ and $B \in \mathcal{B}(\mathbb{R}^d)$). Then $\{X_t : t \in T\}$ is called a *Markov process* if, for every $s, t \in T$ with $s < t$,

$$P[X_t \in B \mid \mathcal{F}_s] = P[X_t \in B \mid \sigma(X_s)] \quad \text{a. s.} \quad \text{for } B \in \mathcal{B}(\mathbb{R}^d), \tag{2.15}$$

where $\sigma(X_s)$ is the σ-algebra generated by X_s.

In Chap. 3 we will encounter Markov processes with $T = \mathbb{R}$. But in this chapter we assume $T = [0, \infty)$ from now on. A readable textbook on conditional probability is Billingsley [13] (2012). We begin with the definition of transition functions.

Definition 2.11 A function $P_{s,t}(x, B)$ with $0 \le s \le t < \infty$, $x \in \mathbb{R}^d$, $B \in \mathcal{B}(\mathbb{R}^d)$ is called *transition function* on $\mathbb{R}^d$ if it is a probability measure with respect to B for fixed s, t, x, a measurable function of x for fixed s, t, B, $P_{s,s}(x, B) = \delta_x(B)$, and

$$\int_{\mathbb{R}^d} P_{s,t}(x, dy) P_{t,u}(y, B) = P_{s,u}(x, B) \qquad \text{for } 0 \le s \le t \le u. \tag{2.16}$$

If, in addition, $P_{s+h,t+h}(x, B)$ does not depend on h, then we write $P_t(x, B) = P_{0,t}(x, B)$ and call it *temporally homogeneous transition function*. In this case (2.16) becomes

$$\int_{\mathbb{R}^d} P_s(x, dy) P_t(y, B) = P_{s+t}(x, B) \qquad \text{for } s \ge 0,\ t \ge 0. \tag{2.17}$$

Definition 2.12 When a transition function $P_{s,t}(x, B)$ is given, a Markov process $\{X_t : t \ge 0\}$ satisfying

$$P[X_t \in B \mid \sigma(X_s)] = P_{s,t}(X_s, B) \quad \text{a. s.} \tag{2.18}$$

for any fixed s, t with $s < t$ and $B \in \mathcal{B}(\mathbb{R}^d)$ is called *Markov process with transition function* $P_{s,t}(x, B)$. When a temporally homogeneous transition function $P_t(x, B)$ is given, a Markov process $\{X_t\}$ satisfying

$$P[X_{s+t} \in B \mid \sigma(X_s)] = P_t(X_s, B) \quad \text{a. s.} \tag{2.19}$$

for any fixed s, t with $s < s + t$ and B is called *temporally homogeneous Markov process with transition function* $P_t(x, B)$.

Proposition 2.13 *Let $\rho \in ID$ and $c \in \mathbb{R}$. Define $P_t(x, B)$ by*

$$\int_{\mathbb{R}^d} e^{i\langle z,y\rangle} P_t(x, dy) = \exp\left[ie^{-ct}\langle x, z\rangle + \int_0^t \psi_\rho(e^{-cs}z)ds\right]. \tag{2.20}$$

Then $P_t(x, B)$ is a temporally homogeneous transition function. It is a distribution in ID for fixed t, x. The process $\{X_t\}$ of Proposition 2.9 is a temporally homogeneous Markov process with transition function $P_t(x, B)$ and initial distribution $\mathcal{L}(J)$.

Proof For every $s \in [0, t]$ we have, from (2.14),

$$X_t = e^{-c(t-s)} X_s + e^{-ct} \int_s^t e^{cu} dZ_u^{(\rho)}. \tag{2.21}$$

Because $e^{-ct} \int_s^t e^{cu} dZ_u^{(\rho)}$ and $\{X_u : u \leq s\}$ are independent, the identity (2.21) shows that $\{X_t : t \geq 0\}$ is a Markov process satisfying (2.15) and (2.18) if we define $P_{s,t}(x, B)$ by

$$P_{s,t}(x, B) = P\left[e^{-c(t-s)} x + e^{-ct} \int_s^t e^{cu} dZ_u^{(\rho)} \in B\right]$$

(use Proposition 1.16 of [93] for a proof). This $P_{s,t}(x, B)$ is infinitely divisible and, by (2.6),

$$\begin{aligned}\int_{\mathbb{R}^d} e^{i\langle z,y\rangle} P_{s,t}(x, dy) &= \exp\left[i\left\langle z, e^{-c(t-s)} x\right\rangle + \int_s^t \psi_\rho\left(e^{-ct} e^{cu} z\right) du\right] \\ &= \exp\left[i\left\langle z, e^{-c(t-s)} x\right\rangle + \int_0^{t-s} \psi_\rho\left(e^{-cv} z\right) dv\right].\end{aligned}$$

Thus $P_{s,t}(x, B)$ depends only on $t - s$. Let $P_t(x, B) = P_{0,t}(x, B)$. Then (2.20) is satisfied. Once the expression is known, it is easy to check (2.17) since, for $\eta(B) = \int P_s(x, dy) P_t(y, B)$, we have

$$\begin{aligned}\widehat{\eta}(w) &= \int P_s(x, dy) \exp\left[ie^{-ct}\langle y, w\rangle + \int_0^t \psi_\rho(e^{-cu} w) du\right] \\ &= \exp\left[ie^{-cs-ct}\langle x, w\rangle + \int_0^s \psi_\rho(e^{-cu-ct} w) du + \int_0^t \psi_\rho(e^{-cu} w) du\right] \\ &= \exp\left[ie^{-cs-ct}\langle x, w\rangle + \int_0^{s+t} \psi_\rho(e^{-cu} w) du\right] = \int P_{s+t}(x, dy) e^{i\langle y,w\rangle}.\end{aligned}$$

It follows that $P_t(x, B)$ is a temporally homogeneous transition function. The assertion is proved. ■

We introduce the following definition.

Definition 2.14 Let $\rho \in ID$ and $c \in \mathbb{R}$. A stochastic process $\{X_t : t \geq 0\}$ on $\mathbb{R}^d$ is called *wide-sense Ornstein–Uhlenbeck type* (or *OU type*) *process* generated by ρ and c, or generated by $\{Z_t^{(\rho)}\}$ and c, if it is a temporally homogeneous Markov process with transition function $P_t(x, B)$ of (2.20). It is called *Ornstein–Uhlenbeck type* (or *OU type*) *process* if, in addition, $c > 0$. The OU type process generated by

Brownian motion and $c > 0$ is called *Ornstein–Uhlenbeck process.*[1] Sometimes the process $\{Z_t^{(\rho)}\}$ is called the *background driving Lévy process*.

Proposition 2.13 shows that from any initial distribution we can construct the wide-sense OU type process generated by ρ and c. An alternative construction is to use the fact that, from any initial distribution and transition function, Kolmogorov extension theorem gives the desired Markov process. The finite-dimensional distributions of the process are expressed by initial distribution and transition function.

Remark 2.15 Let $\rho \in ID$ and $c \in \mathbb{R}$. Let $P_t(x, B)$ be the temporally homogeneous transition function satisfying (2.20). Then ρ and c are determined by $P_t(x, B)$. Indeed, let $P_t'(x, B)$ be defined by

$$\int e^{i\langle z,y\rangle} P_t'(x, dy) = \exp\left[ie^{-c't}\langle x, z\rangle + \int_0^t \psi_{\rho'}(e^{-c's}z)ds\right] \tag{2.22}$$

with some $\rho' \in ID$ and $c' \in \mathbb{R}$. Suppose $P_t(x, B) = P_t'(x, B)$ for all $t,\ x,\ B$. Then

$$ie^{-ct}\langle x, z\rangle + \int_0^t \psi_\rho(e^{-cs}z)ds = ie^{-c't}\langle x, z\rangle + \int_0^t \psi_{\rho'}(e^{-c's}z)ds \tag{2.23}$$

for all z. Differentiation gives

$$-ice^{-ct}\langle x, z\rangle + \psi_\rho(e^{-ct}z) = -ic'e^{-c't}\langle x, z\rangle + \psi_{\rho'}(e^{-c't}z).$$

Letting $t = 0$, we have

$$-ic\langle x, z\rangle + \psi_\rho(z) = -ic'\langle x, z\rangle + \psi_{\rho'}(z). \tag{2.24}$$

Hence, letting $x = 0$, we have $\psi_\rho = \psi_{\rho'}$, that is, $\rho = \rho'$. Hence $c = c'$.

Proposition 2.16 *Suppose that $\{X_t\}$ is a wide-sense OU type process generated by $\rho \in ID$ and $c \in \mathbb{R}$ and, at the same time, by $\rho' \in ID$ and $c' \in \mathbb{R}$. Let $\eta = \mathcal{L}(X_0)$.*

(i) If at least one of ρ and η is non-trivial, then $c = c'$ and $\rho = \rho'$.
(ii) Assume that $\rho = \delta_\gamma$ and $\eta = \delta_{\gamma_\eta}$ for some $\gamma,\ \gamma_\eta \in \mathbb{R}^d$. Then $\rho' = \delta_{\gamma'}$ for some γ' and the following are true.

(1) if $c \neq 0$ and $\gamma/c \neq \gamma_\eta$, then $c' = c$ and $\gamma' = \gamma$;
(2) if $c \neq 0$ and $\gamma/c = \gamma_\eta$, then, either $c' \neq 0$, $\gamma'/c' = \gamma_\eta$ or $c' = 0$, $\gamma' = 0$;
(3) if $c = 0$ and $c' \neq 0$, then $\gamma = 0$ and $\gamma'/c' = \gamma_\eta$;
(4) if $c = 0$ and $c' = 0$, then $\gamma = \gamma'$.
We have $c \neq c'$ or $\gamma \neq \gamma'$ if and only if $\{c, \gamma, c', \gamma'\}$ satisfies one of the following conditions:

[1] Some authors use the name *Ornstein–Uhlenbeck process* for our Ornstein–Uhlenbeck type process.

(5) $c \neq c'$, $c \neq 0$, $c' \neq 0$, $\gamma/c = \gamma'/c' = \gamma_\eta$;
(6) $c \neq 0$, $c' = 0$, $\gamma/c = \gamma_\eta$, $\gamma' = 0$;
(7) $c = 0$, $c' \neq 0$, $\gamma = 0$, $\gamma'/c' = \gamma_\eta$.

Proof The process $\{X_t\}$ is a temporally homogeneous Markov process with transition functions $P_t(x, B)$ and $P'_t(x, B)$ of (2.20) and (2.22).

(i) Assume that at least one of ρ and η is non-trivial. Let $\eta_1 = \mathcal{L}(X_1)$. Then η_1 is non-trivial. We have $P[X_1 \in B_1, X_{1+t} \in B] = \int_{B_1} \eta_1(dx)P_t(x, B) = \int_{B_1} \eta_1(dx)P'_t(x, B)$ for all B_1. Hence, for fixed B and t, $P_t(x, B) = P'_t(x, B)$ for η_1-a. e. x. Since $\mathcal{B}(\mathbb{R}^d)$ is countably generated, we have, for fixed t, $P_t(x, dy) = P'_t(x, dy)$ for η_1-a. e. x. Thus (2.23) holds for all z. It follows from the continuity in t that there is $G \in \mathcal{B}(\mathbb{R}^d)$ with $\eta_1(G) = 1$ such that (2.23) holds for all $x \in G$, t, and z. Hence we have (2.24) for all $x \in G$ and z. That is, $\psi_{\rho'}(z) - \psi_\rho(z) = i(c' - c)\langle x, z\rangle$ for all $x \in G$ and z. Since G is not a singleton, there are distinct points x_1, x_2 in G such that $i(c' - c)\langle x_1, z\rangle = i(c' - c)\langle x_2, z\rangle$ for all z. It follows that $c' = c$. Then we obtain $\rho' = \rho$.

(ii) Since $\rho = \delta_\gamma$, $\int_0^t e^{-cu} dZ_u^{(\rho)}$ equals $(1 - e^{-ct})\gamma/c$ if $c \neq 0$ and equals $t\gamma$ if $c = 0$. Since $\eta = \delta_{\gamma_\eta}$ and

$$P_t(\gamma_\eta, B) = P[e^{-ct}\gamma_\eta + \textstyle\int_0^t e^{-cu} dZ_u^{(\rho)} \in B],$$

$\{X_t\}$ is a non-random process,

$$X_t = \begin{cases} e^{-ct}\gamma_\eta + (1 - e^{-ct})\gamma/c & \text{if } c \neq 0, \\ \gamma_\eta + t\gamma & \text{if } c = 0, \end{cases}$$

$$\frac{dX_t}{dt} = \begin{cases} ce^{-ct}(\gamma/c - \gamma_\eta) & \text{if } c \neq 0, \\ \gamma & \text{if } c = 0. \end{cases}$$

At the same time, these expressions of X_t and dX_t/dt hold with γ and c replaced by γ' and c'. All assertions follow from comparison of the two expressions. ■

Lévy processes do not have limit distributions as $t \to \infty$ except in the case of the zero process. But, in the case of OU type processes, drift force toward the origin of the magnitude proportional to the distance from the origin works, so that they are conjectured to have limit distributions. The conjecture is true only if they do not have too many big jumps.

Theorem 2.17 *Let $c > 0$ be fixed.*

(i) Let $\{Z_t\} = \{Z_t^{(\rho)}\}$ be a Lévy process on $\mathbb{R}^d$ with generating triplet $(A_\rho, \nu_\rho, \gamma_\rho)$, and let $\psi_\rho = \log \widehat{\rho}$. Let $\{X_t\}$ be an OU type process generated by $\{Z_t\}$ and c. Assume that

$$\int_{|x|>2} \log|x|\, \nu_\rho(dx) < \infty. \tag{2.25}$$

Then

$$\mathcal{L}(X_t) \to \mu \quad \text{as } t \to \infty \tag{2.26}$$

for some $\mu \in \mathfrak{P}$, *and this* μ *does not depend on* $\mathcal{L}(X_0)$. *Moreover,*

$$\int_0^\infty \left|\psi_\rho(e^{-cs}z)\right| ds < \infty, \tag{2.27}$$

$$\widehat{\mu}(z) = \exp \int_0^\infty \psi_\rho(e^{-cs}z)ds, \tag{2.28}$$

and μ *belongs to* $L_0(\mathbb{R}^d)$. *The generating triplet* (A, ν, γ) *of* μ *is as follows:*

$$A = (2c)^{-1}A_\rho, \tag{2.29}$$

$$\nu(B) = \int_{\mathbb{R}^d} \nu_\rho(dx) \int_0^\infty 1_B(e^{-cs}x)ds, \quad B \in \mathcal{B}(\mathbb{R}^d), \tag{2.30}$$

$$\gamma = c^{-1}\gamma_\rho + c^{-1} \int_{|x|>1} \frac{x}{|x|} \nu_\rho(dx). \tag{2.31}$$

(ii) For any $\mu \in L_0(\mathbb{R}^d)$, *there exists a unique* $\rho \in ID$ *with generating triplet* $(A_\rho, \nu_\rho, \gamma_\rho)$ *satisfying (2.25) such that* μ *is the limit distribution (2.26) of the OU type process* $\{X_t\}$ *generated by* ρ *and* c *with arbitrary* $\mathcal{L}(X_0)$. *Using* λ *and* $k_\xi(r)$ *in Theorem 1.34 and Remark 1.35 for the Lévy measure* ν *of* μ, *we have*

$$\nu_\rho(B) = -c \int_S \lambda(d\xi) \int_0^\infty 1_B(r\xi)dk_\xi(r). \tag{2.32}$$

(iii) In the set-up of (i), assume

$$\int_{|x|>2} \log|x|\, \nu_\rho(dx) = \infty \tag{2.33}$$

instead of (2.25). Then $\mathcal{L}(X_t)$ *does not tend to any distribution as* $t \to \infty$ *and, moreover, for any* $a > 0$,

$$\sup_{x,y} P_t(x, D_a(y)) \to 0 \quad \text{as } t \to \infty, \tag{2.34}$$

where $D_a(y) = \{z \colon |z - y| \le a\}$.

(Clearly, (2.27), (2.28), (2.30) are rewritten to

$$\int_0^\infty \left|\psi_\rho(e^{-s}z)\right| ds < \infty, \tag{2.35}$$

$$\widehat{\mu}(z) = \exp\left(c^{-1}\int_0^\infty \psi_\rho(e^{-s}z)ds\right), \tag{2.36}$$

$$\nu(B) = c^{-1}\int_{\mathbb{R}^d}\nu_\rho(dx)\int_0^\infty 1_B(e^{-s}x)ds, \quad B\in\mathcal{B}(\mathbb{R}^d), \tag{2.37}$$

respectively. Sometimes we will use this form.)

Proof

(i) From (2.14), the characteristic function of X_t is

$$Ee^{i\langle z,X_t\rangle} = \left(Ee^{ie^{-ct}\langle z,X_0\rangle}\right)\exp\int_0^t \psi_\rho(e^{-cs}z)ds. \tag{2.38}$$

Let $g(z,x) = e^{i\langle z,x\rangle} - 1 - i\langle z,x\rangle 1_{\{|x|\le 1\}}(x)$ as in (1.19). Then

$$g(e^{-cs}z,x) = g(z,e^{-cs}x) + i\langle z,e^{-cs}x\rangle 1_{\{1<|x|\le e^{cs}\}}(x). \tag{2.39}$$

It follows from the Lévy–Khintchine representation (Theorem 1.28) that

$$\int_0^t \psi_\rho(e^{-cs}z)ds = -\tfrac{1}{2}\langle z,\widetilde{A}_t z\rangle + i\langle\widetilde{\gamma}_t,z\rangle + \int_{\mathbb{R}^d} g(z,x)\,\widetilde{\nu}_t(dx), \tag{2.40}$$

$$\widetilde{A}_t = \int_0^t e^{-2cs}ds A_\rho, \tag{2.41}$$

$$\widetilde{\nu}_t(B) = \int_{\mathbb{R}^d}\nu_\rho(dx)\int_0^t 1_B\left(e^{-cs}x\right)ds, \qquad B\in\mathcal{B}\left(\mathbb{R}^d\right), \tag{2.42}$$

$$\widetilde{\gamma}_t = \int_0^t e^{-cs}ds\gamma_\rho + \int_{|x|>1}\nu_\rho(dx)\int_0^t e^{-cs}x 1_{\{|e^{-cs}x|\le 1\}}ds. \tag{2.43}$$

Observe that, as $t\to\infty$, $\widetilde{A}_t$ tends to A of (2.29) and $\widetilde{\nu}_t(B)$ increases to $\nu(B)$ of (2.30). We have $\int(|x|^2\wedge 1)\nu(dx)<\infty$, because

$$\begin{aligned}\int_{|x|\le 1}|x|^2\,\nu(dx) &= \int_{\mathbb{R}^d}\nu_\rho(dx)\int_0^\infty \left|e^{-cs}x\right|^2 1_{\{|e^{-cs}x|\le 1\}}ds\\ &= (2c)^{-1}\int_{\mathbb{R}^d}\left(|x|^2\wedge 1\right)\nu_\rho(dx),\end{aligned}$$

$$\int_{|x|>1}\nu(dx) = \int_{\mathbb{R}^d}\nu_\rho(dx)\int_0^\infty 1_{\{|e^{-cs}x|>1\}}ds = c^{-1}\int_{|x|>1}\log|x|\,\nu_\rho(dx).$$

Since $\widetilde{\nu}_t$ is absolutely continuous with respect to ν with $d\widetilde{\nu}_t/d\nu$ increasing to 1 as $t \to \infty$ and since

$$|g(z,x)| \leq \tfrac{1}{2}|z|^2|x|^2 1_{\{|x|\leq 1\}}(x) + 2\cdot 1_{\{|x|>1\}}(x),$$

we see that

$$\int_{\mathbb{R}^d} g(z,x)\widetilde{\nu}_t(dx) \to \int_{\mathbb{R}^d} g(z,x)\nu(dx) \quad \text{as } t \to \infty.$$

Moreover, since

$$\begin{aligned}\int_{|x|>1} \nu_\rho(dx)\int_0^\infty e^{-cs}|x|1_{\{|x|\leq e^{cs}\}}ds &= \int_{|x|>1}|x|\nu_\rho(dx)\int_{c^{-1}\log|x|}^\infty e^{-cs}ds \\ &= c^{-1}\int_{|x|>1}\nu_\rho(dx),\end{aligned}$$

the dominated convergence theorem gives

$$\begin{aligned}\widetilde{\gamma}_t &\to c^{-1}\gamma_\rho + \int_{|x|>1}\nu_\rho(dx)\int_0^\infty e^{-cs}x1_{\{|x|\leq e^{cs}\}}ds \\ &= c^{-1}\gamma_\rho + c^{-1}\int_{|x|>1}\frac{x}{|x|}\nu_\rho(dx),\end{aligned}$$

which is γ of (2.31). Let μ denote the infinitely divisible distribution with triplet (A,ν,γ). Then we obtain $\exp\int_0^t \psi_\rho(e^{-cs}x)ds \to \widehat{\mu}(z)$ as $t \to \infty$. That is, $\mathcal{L}(X_t) \to \mu$. To show (2.27), notice that

$$\begin{aligned}\left|\psi_\rho(e^{-cs}z)\right| &\leq \tfrac{1}{2}e^{-2cs}\langle z, A_\rho z\rangle + e^{-cs}|\gamma_\rho||z| \\ &\quad + \tfrac{1}{2}|z|^2\int_{|x|\leq e^{cs}}\left|e^{-cs}x\right|^2\nu_\rho(dx) + 2\int_{|x|>e^{cs}}\nu_\rho(dx) \\ &\quad + |z|\int_{1<|x|\leq e^{cs}}\left|e^{-cs}x\right|\nu_\rho(dx),\end{aligned}$$

$$\begin{aligned}\int_0^\infty \left|\psi_\rho(e^{-cs}z)\right|ds &\leq (4c)^{-1}\langle z, A_\rho z\rangle + c^{-1}|\gamma_\rho||z| + \tfrac{1}{2}|z|^2\int_{|x|\leq 1}|x|^2\nu(dx) \\ &\quad + 2\int_{|x|>1}\nu(dx) + c^{-1}|z|\int_{|x|>1}\nu_\rho(dx).\end{aligned}$$

In order to show that $\mu \in L_0$, observe that, for $b > 1$,

$$\widehat{\mu}(b^{-1}z) = \exp\int_0^\infty \psi_\rho(e^{-cs}b^{-1}z)ds = \exp\int_{(\log b)/c}^\infty \psi_\rho(e^{-cs}z)ds.$$

We can write

$$\frac{\widehat{\mu}(z)}{\widehat{\mu}(b^{-1}z)} = \exp \int_0^t \psi_\rho(e^{-cs}z)ds \qquad \text{with } t = (\log b)/c,$$

which is the characteristic function of $P_t(0, dy)$ in Proposition 2.13. Hence $\mu \in L_0$.

(ii) Let $\mu \in L_0$ with generating triplet (A, ν, γ). Then Theorem 1.34 and Remark 1.35 say that

$$\nu(B) = \int_S \lambda(d\xi) \int_0^\infty 1_B(r\xi) \frac{k_\xi(r)}{r} dr, \quad B \in \mathcal{B}(\mathbb{R}^d), \tag{2.44}$$

where λ is a probability measure on S and $k_\xi(r)$ is nonnegative, right continuous, decreasing in $r \in (0, \infty)$, and measurable in $\xi \in S$. Define a measure ν_ρ by (2.32). To prove that ν_ρ is the Lévy measure of some $\rho \in ID$ satisfying (2.25), we will show that

$$\int_{|x| \le 2} |x|^2 \nu_\rho(dx) + \int_{|x|>2} \log |x| \, \nu_\rho(dx) < \infty. \tag{2.45}$$

Let

$$l(u) = \int_0^u (r^2 \wedge 1) \frac{dr}{r} = \begin{cases} (1/2)u^2, & 0 \le u \le 1, \\ 1/2 + \log u, \; u > 1. \end{cases}$$

Below we use the following fact (see [93] Lemma 17.6). In general, if $l(r)$ and $k(r)$ are nonnegative right continuous functions on $(0, \infty)$ such that $k(r)$ is decreasing and $k(\infty) = 0$ and $l(r)$ is increasing and $l(0+) = 0$, then

$$\int_{0+}^\infty l(r) dk(r) = -\int_{0+}^\infty k(r) dl(r), \tag{2.46}$$

including the case of ∞. Now it follows from the definition of l and ν_ρ that

$$\begin{aligned}
\int_{\mathbb{R}^d} l(|x|) \, \nu_\rho(dx) &= -c \int_S \lambda(d\xi) \int_0^\infty l(r) \, dk_\xi(r) \\
&= c \int_S \lambda(d\xi) \int_0^\infty k_\xi(r) dl(r) \\
&= c \int_S \lambda(d\xi) \int_0^\infty (r^2 \wedge 1) \frac{k_\xi(r)}{r} dr \\
&= c \int_{\mathbb{R}^d} (|x|^2 \wedge 1) \nu(dx) < \infty,
\end{aligned}$$

on the one hand. We have

$$\int_{\mathbb{R}^d} l(|x|)\, \nu_\rho(dx) = \int_{|x|\le 1} \frac{1}{2}|x|^2\, \nu_\rho(dx) + \int_{|x|>1} \left(\frac{1}{2} + \log|x|\right) \nu_\rho(dx)$$
$$\ge \int_{|x|\le 2} \frac{1}{8}|x|^2\, \nu_\rho(dx) + \int_{|x|>2} \log|x|\, \nu_\rho(dx)$$

on the other hand. Therefore (2.45) follows.

Now, for $B \in \mathcal{B}(\mathbb{R}^d)$, apply (2.46) to (2.44). Then

$$\nu(B) = -\int_S \lambda(d\xi) \int_0^\infty dk_\xi(r) \int_0^r 1_B(u\xi)\frac{du}{u}$$
$$= -\int_S \lambda(d\xi) \int_0^\infty dk_\xi(r) \int_0^\infty 1_B(e^{-s}r\xi)ds$$
$$= c^{-1} \int_{\mathbb{R}^d} \nu_\rho(dy) \int_0^\infty 1_B(e^{-s}y)ds,$$

including the case of ∞. That is, (2.30) holds.

Next, define A_ρ and γ_ρ by (2.29) and (2.31), respectively. Then, the OU type process generated by ρ with generating triplet $(A_\rho, \nu_\rho, \gamma_\rho)$ and c has μ as limit distribution. This proves (2.26).

We next show the uniqueness of the generating triplet. Suppose that two processes of OU type with common c have the limit distribution μ. Let $\{Z_t^1\}$ and $\{Z_t^2\}$ be their background driving Lévy processes with $\rho_j = \mathcal{L}(Z_1^j)$ and $\psi_j = \log \widehat{\rho}_j$, $j = 1, 2$. By (2.28) we have $\int_0^\infty \psi_1(e^{-cs}z)ds = \int_0^\infty \psi_2(e^{-cs}z)ds$. Replacing z by $e^{-ct}z$, we obtain $\int_t^\infty \psi_1(e^{-cs}z)ds = \int_t^\infty \psi_2(e^{-cs}z)ds$. Hence $\int_0^t \psi_1(e^{-cs}z)ds = \int_0^t \psi_2(e^{-cs}z)ds$. Differentiation at $t = 0$ leads to $\psi_1(z) = \psi_2(z)$.

(iii) We assume (2.33). Let $\widetilde{\nu}_t$ be the Lévy measure of $P_t(x, dy)$; $\widetilde{\nu}_t$ does not depend on x. Then, Proposition 2.13 and (2.42) give

$$\int_{|x|>a} \widetilde{\nu}_t(dx) = \int_{|x|>a} \left(t \wedge \left(\frac{1}{c}\log\frac{|x|}{a}\right)\right) \nu_\rho(dx) \to \infty \quad \text{as } t \to \infty \tag{2.47}$$

for any $a > 0$. It follows that, for any $x \in \mathbb{R}^d$, $P_t(x, dy)$ does not tend to a probability measure as $t \to \infty$. Indeed, if on the contrary $P_t(x_0, dy)$ tends to some probability measure μ for some x_0, then, μ is infinitely divisible (Proposition 1.5) and its Lévy measure ν satisfies that, for any bounded continuous function f vanishing on a neighbourhood of 0, $\int f(x)\widetilde{\nu}_t(dx) \to \int f(x)\,\nu(dx)$ (see [93] Theorem 8.7), which contradicts (2.47). Since (2.20) and (2.38) give

$$Ee^{i\langle z, X_t\rangle} = \left(Ee^{i\langle z, e^{-ct}X_0\rangle}\right) \int e^{i\langle x,y\rangle} P_t(0, dy),$$

it follows that $\mathcal{L}(X_t)$ does not tend to any probability measure as $t \to \infty$. We use (2.47) and Lemma 2.18 given below to obtain

$$P_t(x, D_a(y)) \leq C_d \left(\int_{|u|>a/\pi} \widetilde{\nu}_t(du) \right)^{-1/2} \to 0 \quad as \quad t \to \infty,$$

for any a, x, and y, with a constant C_d depending only on d. Thus we get the assertion (2.34). This finishes the proof of (iii), provided that Lemma 2.18 is true. ■

The following lemma is an interesting estimate of $\mu(B)$ for $\mu \in ID$ and a bounded Borel set B by the tail of the Lévy measure ν of μ. It is written in Hengartner and Theodorescu [33] (1973) for $d = 1$ and called *LeCam's estimate*; it is extended to general d by Sato and Yamazato [110] (1984).

Lemma 2.18 *Let $I(x, a) = [x_1 - a, x_1 + a] \times \cdots \times [x_d - a, x_d + a]$, a cube in $\mathbb{R}^d$ with centre $x = (x_j)_{1 \leq j \leq d}$. Let $\mu \in ID(\mathbb{R}^d)$ with Lévy measure ν. Then*

$$\mu(I(x, a)) \leq C_d \left(\int_{|y|>a/\pi} \nu(dy) \right)^{-1/2}, \tag{2.48}$$

where C_d is a constant which depends only on d.

Proof First we show that, for any $\mu \in \mathfrak{P}(\mathbb{R}^d)$, $x \in \mathbb{R}^d$, $a > 0$, and $b > 0$ with $b \leq \pi/a$,

$$\mu\left(I(x, a)\right) \leq \left(\frac{\pi}{2}\right)^{2d} b^{-d} \int_{I(0,b)} |\widehat{\mu}(z)|\, dz. \tag{2.49}$$

Let $f(u) = \left(\frac{\sin(u/2)}{u/2}\right)^2$ and $h(v) = (1 - |v|)1_{\{|v| \leq 1\}}(v)$. Then

$$f(u) = \int_{-\infty}^{\infty} e^{iuv} h(v) dv, \qquad h(v) = \frac{1}{2\pi} \int_{-\infty}^{\infty} e^{-iuv} f(u) du.$$

For $x, z \in \mathbb{R}^d$, let $\widetilde{f}(x) = \prod_{j=1}^d f(x_j)$ and $\widetilde{h}(z) = \prod_{j=1}^d h(z_j)$. Then, for every $x \in \mathbb{R}^d$ and $b > 0$,

$$\begin{aligned}\int \widetilde{f}(b(y - x))\mu(dy) &= \int \mu(dy) \int e^{i\langle b(y-x), z\rangle} \prod_{j=1}^{d} h(z_j) dz \\ &= b^{-d} \int e^{-i\langle x, z\rangle} \widehat{\mu}(z) \widetilde{h}(b^{-1} z) dz.\end{aligned}$$

Since $f(u) \geq (2/\pi)^2$ for $|u| \leq \pi$, it follows that

$$b^{-d}\int_{I(0,b)}|\widehat{\mu}(z)|\,dz \geq \int_{I(x,a)}\widetilde{f}(b(y-x))\mu(dy) \geq \left(\frac{2}{\pi}\right)^{2d}\mu\left(I(x,a)\right)$$

if $ab \leq \pi$, that is, (2.49).

Now let $\mu \in ID$ with Lévy measure ν. We claim that, for any $b > 0$,

$$b^{-d}\int_{I(0,b)}|\widehat{\mu}(z)|\,dz \leq C'_d\left(\int_{|y|>1/b}\nu(dy)\right)^{-1/2}, \tag{2.50}$$

where C'_d is a constant which depends only on d. We have

$$|\widehat{\mu}(z)| \leq \exp\left[\mathrm{Re}\int g(z,y)\nu(dy)\right] \leq \exp\left[-\int_{|y|>1/b}(1-\cos\langle z,y\rangle)\nu(dy)\right].$$

Let $V = \int_{|y|>1/b}\nu(dy)$. If $V = 0$, then (2.50) is trivial. Suppose that $V > 0$, and let $\widetilde{\nu}(dy) = V^{-1}1_{\{|y|>1/b\}}(y)\nu(dy)$. Since

$$|\widehat{\mu}(z)| \leq \int e^{-V(1-\cos\langle z,y\rangle)}\widetilde{\nu}(dy)$$

by Jensen's inequality for the convex function e^{-u}, we have

$$\int_{I(0,b)}|\widehat{\mu}(z)|\,dz \leq \int_{|y|>1/b}F(y)\widetilde{\nu}(dy) \text{ with } F(y) = \int_{|z|\leq\sqrt{d}b}e^{-V(1-\cos\langle z,y\rangle)}dz.$$

We fix $y \neq 0$ and consider an orthogonal transformation that carries $y/|y|$ to $e_1 = (\delta_{1j})_{1\leq j\leq d}$. Then

$$F(y) = \int_{|z|\leq\sqrt{d}b}e^{-V(1-\cos(z_1|y|))}dz.$$

Let $E_k = \{z \in \mathbb{R}^d\colon |z| \leq \sqrt{d}b \text{ and } 2\pi k/|y| < z_1 \leq 2\pi(k+1)/|y|\}$ and $n = \left[\sqrt{d}b|y|/2\pi\right]$ with brackets denoting the integer part. Then

$$\begin{aligned}F(y) &= 2\sum_{k=0}^{n}\int_{E_k}e^{-V(1-\cos(z_1|y|))}dz\\ &\leq 2(n+1)\int\cdots\int_E dz_2\cdots dz_d\int_0^{2\pi/|y|}e^{-V(1-\cos(z_1|y|))}dz_1\\ &\leq 4C''_d b^{d-1}(n+1)|y|^{-1}\int_0^{\pi}e^{-V(1-\cos u)}du,\end{aligned}$$

where $E = \{z' \in \mathbb{R}^{d-1} \colon |z'| \leq \sqrt{d}b\}$ and C''_d is the volume of the ball with radius $\sqrt{d}$ in $\mathbb{R}^{d-1}$. Using $1 - \cos u \geq 2\pi^{-2}u^2$ for $0 \leq u \leq \pi$, we have

$$\int_0^{\pi} e^{-V(1-\cos u)} du \leq \int_0^{\infty} e^{-2V\pi^{-2}u^2} du = CV^{-1/2}$$

with an absolute constant C. Noting that

$$\sup_{|y|>1/b} (n+1)|y|^{-1} = \sup_{|y|>1/b} \left(\left[\sqrt{d}b|y|/2\pi \right] + 1 \right) |y|^{-1} = bC'''_d$$

with a constant C'''_d depending only on d, we obtain (2.50).

Taking $b = \pi/a$ and combining (2.49) and (2.50), we get (2.48). ∎

Remark 2.19 In Theorem 2.17 (i), $\mathcal{L}(X_t)$ converges as $t \to \infty$. But, if $\{Z_t\} = \{Z_t^{(\rho)}\}$ is non-trivial, then X_t does not converge in probability as $t \to \infty$. Proof is as follows. It is enough to consider the case $X_t = e^{-ct} \int_0^t e^{cs} dZ_s$. Suppose, on the contrary, that $X_t \to Y$ in probability for some Y as $t \to \infty$. Then $X_t - X_{t-1} \to 0$ in probability, since

$$P[|X_t - X_{t-1}| > \varepsilon] \leq P[|X_t - Y| > \varepsilon/2] + P[|X_{t-1} - Y| > \varepsilon/2] \to 0$$

for any $\varepsilon > 0$. Hence $Ee^{i\langle z, X_t - X_{t-1}\rangle} \to 1$ as $t \to \infty$. From the non-triviality there is $z_0 \in \mathbb{R}^d$ such that $|\widehat{\rho}(z_0)| < 1$ (see [93] Lemma 13.9). It follows that $\operatorname{Re} \psi_\rho(z_0) < 0$ since $|\widehat{\rho}(z_0)| = e^{\operatorname{Re} \psi_\rho(z_0)}$. As $\operatorname{Re} \psi_\rho \leq 0$ and ψ_ρ is continuous, it follows that $\int_0^1 \operatorname{Re} \psi_\rho(e^{-cs} z_0) ds < 0$. We have

$$X_t - X_{t-1} = e^{-ct} \int_{t-1}^{t} e^{cs} dZ_s + (e^{-ct} - e^{-c(t-1)}) \int_0^{t-1} e^{cs} dZ_s$$

and the two terms on the right are independent. Hence

$$\begin{aligned} \left| Ee^{i\langle z, X_t - X_{t-1}\rangle} \right| &\leq \left| E \exp \left[i\langle z, e^{-ct} \int_{t-1}^{t} e^{cs} dZ_s \rangle \right] \right| = \left| \exp \int_{t-1}^{t} \psi_\rho(e^{-ct+cs} z) ds \right| \\ &= \left| \exp \int_0^1 \psi_\rho(e^{-cs} z) ds \right| = \exp \int_0^1 \operatorname{Re} \psi_\rho(e^{-cs} z) ds < 1, \end{aligned}$$

if $z = z_0$. This is a contradiction.

It is natural to consider Theorem 2.17 as a result on transformation of ρ to μ. Thus we give the following two definitions and one remark.

Definition 2.20 The class $ID_{\log} = ID_{\log}(\mathbb{R}^d)$ is the collection of $\rho \in ID(\mathbb{R}^d)$ such that its Lévy measure ν_ρ satisfies $\int_{|x|>2} \log |x| \nu_\rho(dx) < \infty$ or, equivalently, $\int_{|x|>2} \log |x| \rho(dx) < \infty$; see [93] Theorem 25.3 and Proposition 25.4.

Definition 2.21 Let $c > 0$. For $\rho \in ID(\mathbb{R}^d)$ define a mapping $\Phi_{(c)}(\rho) = \mu$ if the OU type process $\{X_t\}$ generated by ρ and c, starting from an arbitrary initial distribution, satisfies $\mathcal{L}(X_t) \to \mu$ as $t \to \infty$. Let $\mathfrak{D}(\Phi_{(c)})$ and $\mathfrak{R}(\Phi_{(c)})$, respectively, denote the domain (of definition) and the range of $\Phi_{(c)}$. That is, $\mathfrak{R}(\Phi_{(c)}) = \{\Phi_{(c)}(\rho)\colon \rho \in \mathfrak{D}(\Phi_{(c)})\}$.

Remark 2.22 Theorem 2.17 says that $\mathfrak{D}(\Phi_{(c)}) = ID_{\log}$, $\mathfrak{R}(\Phi_{(c)}) = L_0$, $\Phi_{(c)}$ is one-to-one, and $\Phi_{(c)}(\rho)$ does not depend on X_0 . The mapping $\Phi_{(c)}$ depends on c, but neither $\mathfrak{D}(\Phi_{(c)})$ nor $\mathfrak{R}(\Phi_{(c)})$ depends on c.

In general we write $\psi_\mu = \log \widehat{\mu}$ for $\mu \in ID$.

Proposition 2.23 *Let ρ, $\rho' \in ID$ and c, $c' > 0$.*

(i) If $\mu = \Phi_{(c)}(\rho)$, then $\psi_\mu(z) = \int_0^\infty \psi_\rho(e^{-cs}z)ds$.
(ii) $\Phi_{(c)}(\rho * \rho') = \Phi_{(c)}(\rho) * \Phi_{(c)}(\rho')$.
(iii) $\Phi_{(c)}(\rho^{t*}) = (\Phi_{(c)}(\rho))^{t*}$ for $t \geq 0$.
(iv) $\Phi_{(c')}(\rho) = (\Phi_{(c)}(\rho))^{(c/c')*}$.
(v) If $\Phi_{(c)}(\rho) = \Phi_{(c')}(\rho')$, then $\rho' = \rho^{(c'/c)*}$.
(vi) If $\rho = \delta_{\gamma_\rho}$, then $\Phi_{(c)}(\rho) = \delta_{\gamma_\rho/c}$.

Proof

(i) This is shown in Theorem 2.17.
(ii) Let $\mu = \Phi_{(c)}(\rho)$, $\mu' = \Phi_{(c)}(\rho')$, and $\mu'' = \Phi_{(c)}(\rho * \rho')$. Then $\psi_{\mu''}(z) = \int_0^\infty \psi_{\rho*\rho'}(e^{-cs}z)ds = \int_0^\infty(\psi_\rho(e^{-cs}z) + \psi_{\rho'}(e^{-cs}z))ds = \psi_\mu + \psi_{\mu'}(z)$.
(iii) Let $\mu = \Phi_{(c)}(\rho)$ and $\mu_t = \Phi_{(c)}(\rho^{t*})$. Then $\psi_{\mu_t}(z) = \int_0^\infty \psi_{\rho^{t*}}(e^{-cs}z)ds = t\int_0^\infty \psi_\rho(e^{-cs}z)ds$.
(iv) Let $\mu_c = \Phi_{(c)}(\rho)$ and $\mu_{c'} = \Phi_{(c')}(\rho)$. Then $\psi_{\mu_c}(z) = \int_0^\infty \psi_\rho(e^{-cs}z)ds = (1/c)\int_0^\infty \psi_\rho(e^{-s}z)ds$ and $\psi_{\mu_{c'}}(z) = (1/c')\int_0^\infty \psi_\rho(e^{-s}z)ds$. Hence $\psi_{\mu_{c'}}(z) = (c/c')\psi_{\mu_c}(z)$.
(v) We have $\Phi_{(c')}(\rho^{(c'/c)*}) = (\Phi_{(c')}(\rho))^{(c'/c)*} = \Phi_{(c)}(\rho)$, using (iv). Hence $\Phi_{(c')}(\rho^{(c'/c)*}) = \Phi_{(c')}(\rho')$. Then use the one-to-one property.
(vi) If $\rho = \delta_{\gamma_\rho}$, then $\psi_\rho(z) = i\langle\gamma_\rho, z\rangle$ and $\int_0^\infty \psi_\rho(e^{-cs}z)ds = i\langle\gamma_\rho, z\rangle/c$. ∎

Another equivalent definition of $\Phi_{(c)}$ is by the distribution of an improper stochastic integral.

Definition 2.24 If the limit in probability of $\int_{t_0}^t f(s)dZ_s^{(\rho)}$ as $t \to \infty$ exists, then the limit is denoted by $\int_{t_0}^\infty f(s)dZ_s^{(\rho)}$ and we say that $\int_{t_0}^\infty f(s)dZ_s^{(\rho)}$ is *definable*. The limit is called the *improper stochastic integral* of f.

Proposition 2.25 *Let $f(s)$ be measurable and locally bounded on $[0, \infty)$ (that is, bounded on every bounded interval $[0, t]$). Then the improper stochastic integral $\int_0^\infty f(s)dZ_s^{(\rho)}$ is definable if and only if $\int_0^t \psi_\rho(f(s)z)ds$ is convergent in $\mathbb{C}$ as $t \to \infty$ for each $z \in \mathbb{R}^d$. In this case*

$$Ee^{i\langle z,X\rangle} = \exp\left(\lim_{t\to\infty}\int_0^t \psi_\rho(f(s)z)ds\right) \quad \textit{for } X = \int_0^\infty f(s)dZ_s^{(\rho)}. \tag{2.51}$$

Proof The "only if" part follows from Propositions 2.2 and 2.4, since convergence in probability implies convergence in distribution. To see the "if" part, notice that

$$E \exp\left(i\left\langle z, \int_{t_0}^{t_1} f(s)dZ_s^{(\rho)}\right\rangle\right) = \exp\int_{t_0}^{t_1} \psi_\rho(f(s)z)ds \to 1, \quad t_0, t_1 \to \infty.$$

It follows that, for every $\varepsilon > 0$, $P\left(\left|\int_{t_0}^{t_1} f(s)dZ_s^{(\rho)}\right| > \varepsilon\right) \to 0$ as $t_0, t_1 \to \infty$. Hence $\int_0^t f(s)dZ_s^{(\rho)}$ converges in probability as $t \to \infty$. ■

Definition 2.26 In the set-up of Proposition 2.25, the *improper stochastic integral mapping* Φ_f is defined as follows:

$$\Phi_f(\rho) = \mathcal{L}\left(\int_0^\infty f(s)dZ_s^{(\rho)}\right),$$

$$\mathfrak{D}(\Phi_f) = \left\{\rho \in ID\colon \int_0^\infty f(s)dZ_s^{(\rho)} \text{ is definable}\right\}.$$

Lemma 2.27 *Let $\{Y_t : t \geq 0\}$ be an additive process in law on $\mathbb{R}^d$. Then, $Y_t \to Y_\infty$ in probability for some Y_∞ as $t \to \infty$ if and only if $\mathcal{L}(Y_t) \to \mu$ for some $\mu \in \mathfrak{P}$ as $t \to \infty$. In this case, $\mathcal{L}(Y_\infty) = \mu$.*

Proof The "only if" part is well-known. Let us prove the "if" part. Suppose that $\mathcal{L}(Y_t) \to \mu$. Let $\rho_{s,t} = \mathcal{L}(Y_t - Y_s)$ for $0 \leq s \leq t < \infty$. Then $\rho_{0,t} = \rho_{0,s} * \rho_{s,t}$, $\rho_{s,t} \in ID$, and $\mu \in ID$. Their characteristic functions have no zero. Hence $\widehat{\rho}_{s,t}(z) = \widehat{\rho}_{0,t}(z)/\widehat{\rho}_{0,s}(z) \to \widehat{\mu}(z)/\widehat{\mu}(z) = 1 = \widehat{\delta}_0(z)$ as $s, t \to \infty$. Thus, for any $\varepsilon > 0$, $P[|Y_t - Y_s| > \varepsilon] \to 0$ as $s, t \to \infty$. This implies that Y_t is convergent in probability ([19] Exercise 4.2.6). ■

Proposition 2.28 *Fix $c > 0$. Let $\Phi_{(c)}$ be as defined in Definition 2.21. If $f(s) = e^{-cs}$, then $\Phi_f = \Phi_{(c)}$, which means that $\mathfrak{D}(\Phi_f) = \mathfrak{D}(\Phi_{(c)})$ and that $\Phi_f(\rho) = \Phi_{(c)}(\rho)$ for $\rho \in \mathfrak{D}(\Phi_f)$. We also have*

$$\mathfrak{D}(\Phi_{(c)}) = \left\{\rho \in ID\colon \int_0^\infty |\psi_\rho(e^{-cs}z)|ds < \infty \textit{ for } z \in \mathbb{C}\right\}. \tag{2.52}$$

Proof Let $Y_t = \int_0^t e^{-cs}dZ_s^{(\rho)}$. Then $\{Y_t\}$ is an additive process in law on $\mathbb{R}^d$ and $\mathcal{L}(Y_t)$ has characteristic function $\exp\int_0^t \psi_\rho(e^{-cs}z)ds$ (Proposition 2.4). Hence $\mathcal{L}(Y_t)$ equals $P_t(0, dy)$ (Proposition 2.13). That is, $\mathcal{L}(Y_t) = \mathcal{L}(X_t)$ whenever $J = 0$. Thus we obtain $\Phi_f = \Phi_{(c)}$ for $f(s) = e^{-cs}$ by Theorem 2.17 and Lemma 2.27. To see (2.52), let $\mathfrak{M}$ be the right-hand side of (2.52). Then $\mathfrak{D}(\Phi_{(c)}) = ID_{\log} \subset \mathfrak{M}$ from Theorem 2.17. If $\rho \in \mathfrak{M}$, then $\int_0^t \psi_\rho(e^{-cs}z)ds$ is convergent as $t \to \infty$ for each z, and hence $\int_0^\infty f(s)dZ_s$ is definable by Proposition 2.25. Therefore $\mathfrak{D}(\Phi_{(c)}) \supset \mathfrak{M}$. ■

Incidentally, for some $f(s)$ decreasing to 0 as $s \to \infty$, we have

$$\mathfrak{D}(\Phi_f) \supsetneqq \{\rho \in ID\colon \textstyle\int_0^\infty |\psi_\rho(f(s)z)|ds < \infty \text{ for } z \in \mathbb{C}\}$$

(see Sato [101] (1984)).

2.3 Relations to Classes L_m, $\mathfrak{S}_\alpha$, and $\mathfrak{S}_\alpha^0$

We clarify the relation of the mapping $\Phi_{(c)}$ in Definition 2.21 with the classes L_m, $\mathfrak{S}$, $\mathfrak{S}^0$, $\mathfrak{S}_\alpha$, and $\mathfrak{S}_\alpha^0$ introduced in Sect. 1.1.

Theorem 2.29 *Let $c > 0$.*

(i) Let $m \in \{0, 1, \ldots, \infty\}$. Then $\Phi_{(c)}$ maps $L_{m-1} \cap ID_{\log}$ onto L_m one-to-one. Here we understand $L_{-1} = ID$.

(ii) Let $0 < \alpha \le 2$. Then, $\rho \in \mathfrak{S}_\alpha$ if and only if $\rho \in ID_{\log}$ and $\Phi_{(c)}(\rho) \in \mathfrak{S}_\alpha$. Further, $\rho \in \mathfrak{S}_\alpha^0$ if and only if $\rho \in ID_{\log}$ and $\Phi_{(c)}(\rho) \in \mathfrak{S}_\alpha^0$.

(iii) If $\rho \in \mathfrak{S}_\alpha$, then $\rho \in ID_{\log}$ and

$$\Phi_{(c)}(\rho) = \rho^{1/(\alpha c)*} * \delta_\gamma \qquad \text{for some } \gamma \in \mathbb{R}^d. \tag{2.53}$$

Conversely, if ρ is a non-trivial distribution in $ID_{\log}$ and

$$\Phi_{(c)}(\rho) = \rho^{t*} * \delta_\gamma \qquad \text{for some } t > 0 \text{ and } \gamma \in \mathbb{R}^d, \tag{2.54}$$

then $1/(tc) \le 2$ and $\rho \in \mathfrak{S}_{1/(tc)}$. The above two sentences remain true with $\mathfrak{S}_\alpha$, $\mathfrak{S}_{1/(tc)}$, and γ replaced by $\mathfrak{S}_\alpha^0$, $\mathfrak{S}_{1/(tc)}^0$, and 0, respectively.

Proof

(i) Let $\rho \in ID_{\log}$ and $\Phi_{(c)}(\rho) = \mu$. Then $\mu \in L_0$ and, for each $b > 1$, there is $\eta_b \in ID$ such that $\widehat{\mu}(z) = \widehat{\mu}(b^{-1}z)\widehat{\eta}_b(z)$. This η_b satisfies

$$\widehat{\eta}_b(z) = \exp \int_0^{(1/c)\log b} \psi_\rho(e^{-cs}z)ds, \tag{2.55}$$

as is seen at the end of the proof of Theorem 2.17 (i). If $\mu \in L_m$, then $\eta_b \in L_{m-1}$ and

$$\widehat{\eta_b^{c/\log b}}(z) = \exp\left[\frac{c}{\log b}\int_0^{(1/c)\log b} \psi_\rho(e^{-cs}z)ds\right] = \exp \int_0^1 \psi_\rho(b^{-u}z)du$$

$$\to \exp \psi_\rho(z) = \widehat{\rho}(z) \quad \text{as } b \downarrow 1,$$

proving that $\rho \in L_{m-1}$ by the use of Proposition 1.18. Conversely, let $\rho \in L_{m-1} \cap ID_{\log}$. Recall that the right-hand side of (2.55) is the characteristic function of $\int_0^{(1/c)\log b} e^{-cs} dZ_s^{(\rho)}$. If $f(s)$ is a step function, then $\mathcal{L}(\int_0^t f(s) dZ_s^{(\rho)}) \in L_{m-1}$, see (2.4). Then, it follows from Propositions 1.18, 2.2, and 2.4 that $\eta_b \in L_{m-1}$. Therefore $\mu \in L_m$. Now we see that $\Phi_{(c)}$ maps L_{m-1} onto L_m. The one-to-one property is known in Theorem 2.17 (ii).

(ii) By Definition 1.23, a distribution ρ in ID is α-stable if and only if, for any $t > 0$, there is $\gamma_{\rho,t} \in \mathbb{R}^d$ such that

$$t\psi_\rho(z) = \psi_\rho(t^{1/\alpha} z) + i\langle \gamma_{\rho,t}, z \rangle. \tag{2.56}$$

We note that any stable distribution is in $ID_{\log}$, which follows from (1.40). If $\rho \in \mathfrak{S}_\alpha$ and $\Phi_{(c)}(\rho) = \mu$, then, by (2.28),

$$\begin{aligned}\widehat{\mu}(z)^t &= \exp\left[t \int_0^\infty \psi_\rho(e^{-cs} z) ds\right] \\ &= \exp \int_0^\infty \left(\psi_\rho(t^{1/\alpha} e^{-cs} z) + i\langle \gamma_{0,t}, e^{-cs} z \rangle\right) ds = \widehat{\mu}(t^{1/\alpha} z) e^{i\langle \gamma_t, z \rangle}\end{aligned}$$

with $\gamma_t = (1/c)\gamma_{0,t}$, which shows that $\mu \in \mathfrak{S}_\alpha$.

Conversely, assume that $\mu \in \mathfrak{S}_\alpha$, $\rho \in ID_{\log}$, and $\mu = \Phi_{(c)}(\rho)$. Then $\widehat{\mu}(z)^t = \widehat{\mu}(t^{1/\alpha} z) e^{i\langle \gamma_t, z \rangle}$ with some γ_t, and hence, by (2.28),

$$t \int_0^\infty \psi_\rho(e^{-cs} z) ds = \int_0^\infty \psi_\rho(e^{-cs} t^{1/\alpha} z) ds + i\langle \gamma_t, z \rangle$$

for all $z \in \mathbb{R}^d$. Replacing z by $e^{-cu} z$ and making change of variables, we get

$$t \int_u^\infty \psi_\rho(e^{-cs} z) ds = \int_u^\infty \psi_\rho(e^{-cs} t^{1/\alpha} z) ds + i\langle e^{-cu} \gamma_t, z \rangle.$$

Differentiation in u gives

$$t\psi_\rho(e^{-cu} z) = \psi_\rho(e^{-cu} t^{1/\alpha} z) + i\langle c e^{-cu} \gamma_t, z \rangle.$$

Letting $u \downarrow 0$, we have $t\psi_\rho(z) = \psi_\rho(t^{1/\alpha} z) + i\langle c\gamma_t, z \rangle$, that is, $\rho \in \mathfrak{S}_\alpha$. The argument above simultaneously shows that $\rho \in \mathfrak{S}^0_\alpha$ if and only if $\mu \in \mathfrak{S}^0_\alpha$.

(iii) Let $\rho \in \mathfrak{S}_\alpha \subset ID_{\log}$. We will show (2.53). Since (2.56) holds for all $t > 0$, we have

$$\int_0^\infty \psi_\rho(e^{-cs} z) ds = \int_0^\infty \left(e^{-\alpha cs} \psi_\rho(z) - i\langle \gamma_{0,\exp(-\alpha cs)}, z \rangle\right) ds.$$

It follows that

$$\int_0^\infty \psi_\rho(e^{-cs}z)ds = \frac{1}{\alpha c}\psi_\rho(z) + i\langle \gamma, z\rangle$$

with $\gamma = -\lim_{u\to\infty}\int_0^u \gamma_{0,\exp(-\alpha cs)}ds$, where the existence of the limit comes from the finiteness of $\int_0^\infty \psi_\rho(e^{-cs}z)ds$ and $\int_0^\infty e^{-\alpha cs}\psi_\rho(z)ds$. By (2.28) this gives (2.53).

Conversely, suppose that ρ is non-trivial, belongs to $ID_{\log}$, and satisfies (2.54). This means that

$$\int_0^\infty \psi_\rho(e^{-cs}z)ds = t\psi_\rho(z) + i\langle \gamma, z\rangle.$$

Let $u \in \mathbb{R}$ and replace z by $e^{-cu}z$ to obtain

$$\int_u^\infty \psi_\rho(e^{-cv}z)dv = t\psi_\rho(e^{-cu}z) + i\langle e^{-cu}\gamma, z\rangle.$$

Fix z for a while and denote $f(u) = i\langle ce^{-cu}\gamma, z\rangle$ and $g(u) = \psi_\rho(e^{-cu}z)$ for $u \in \mathbb{R}$. Then we see that g is differentiable and $g(u) = -tg'(u) + f(u)$. Hence

$$\frac{d}{dt}\left(e^{u/t}g(u)\right) = e^{u/t}g'(u) + \frac{1}{t}e^{u/t}g(u) = \frac{1}{t}e^{u/t}f(u),$$

that is,

$$e^{u/t}g(u) = \int_0^u \frac{1}{t}e^{v/t}f(v)dv + g(0) \quad \text{for } u \in \mathbb{R}.$$

Now we have

$$\begin{aligned}\psi_\rho(e^{-cu}z) &= e^{-u/t}\int_0^u \frac{1}{t}e^{v/t}i\langle ce^{-cv}\gamma, z\rangle dv + e^{-u/t}\psi_\rho(z)\\ &= i\langle \eta_u, z\rangle + e^{-u/t}\psi_\rho(z)\end{aligned}$$

with some $\eta_u \in \mathbb{R}^d$. Thus, for every $s > 0$, there is $\gamma_{0,s}$ such that

$$s\psi_\rho(z) = \psi_\rho(s^{tc}z) + i\langle \gamma_{0,s}, z\rangle \quad \text{for } z \in \mathbb{R}^d.$$

Since ρ is non-trivial, this shows that $(tc)^{-1} \le 2$ and ρ is $(tc)^{-1}$-stable by Propositions 1.21 and 1.22.

The argument above also proves the last assertion in (iii) concerning strict stability. ∎

Remark 2.30 If ρ is stable and $\mu = \Phi_{(c)}(\rho)$, then there are $a > 0$ and $\gamma \in \mathbb{R}^d$ satisfying

$$\widehat{\mu}(z) = \widehat{\rho}(az)e^{i\langle\gamma,z\rangle}.$$

But the converse is not true. See Wolfe [139] (1982a).

Remark 2.31 By introducing a stronger convergence concept in $ID_{\log}$ and using the usual weak convergence in L_0, the mapping $\Phi_{(c)}$ and its inverse are continuous. See Sato and Yamazato [110] (1984).

Let us give another formulation of the relation of $\Phi_{(c)}$ with the classes L_m. For $m \in \mathbb{N}$, let $\Phi^m_{(c)}$ be the mth iteration of $\Phi_{(c)}$. That is, $\Phi^1_{(c)} = \Phi_{(c)}$ and, for $m \geq 2$, $\Phi^m_{(c)}$ is defined by $\Phi^m_{(c)}(\rho) = \Phi_{(c)}(\Phi^{m-1}_{(c)}(\rho))$ if $\rho \in \mathfrak{D}(\Phi^{m-1}_{(c)})$ and $\Phi^{m-1}_{(c)}(\rho) \in \mathfrak{D}(\Phi_{(c)})$. The domain and the range of $\Phi^m_{(c)}$, $\mathfrak{D}(\Phi^m_{(c)})$ and $\mathfrak{R}(\Phi^m_{(c)})$, are described below.

Theorem 2.32 *Fix $c > 0$ and let m be a positive integer. For μ and ρ in $ID(\mathbb{R}^d)$ we write $\psi = \log\widehat{\mu}$, $\psi_\rho = \log\widehat{\rho}$, (A, ν, γ) for the triplet of μ, and $(A_\rho, \nu_\rho, \gamma_\rho)$ for the triplet of ρ. Then,*

$$\mathfrak{D}(\Phi^m_{(c)}) = \left\{\rho \in ID\colon \textstyle\int_{|x|>2}(\log|x|)^m \nu_\rho(dx) < \infty\right\}, \tag{2.57}$$

$$\mathfrak{R}(\Phi^m_{(c)}) = L_{m-1}. \tag{2.58}$$

If $\rho \in \mathfrak{D}(\Phi^m_{(c)})$ and $\mu = \Phi^m_{(c)}(\rho)$, then

$$\int_0^\infty s^{m-1}\left|\psi_\rho(e^{-cs}z)\right| ds < \infty, \tag{2.59}$$

$$\psi(z) = \int_0^\infty \frac{s^{m-1}}{(m-1)!}\psi_\rho(e^{-cs}z)ds. \tag{2.60}$$

$$A = (2c)^{-m}A_\rho, \tag{2.61}$$

$$\nu(B) = \int_{\mathbb{R}^d}\nu_\rho(dx)\int_0^\infty \frac{s^{m-1}}{(m-1)!}1_B(e^{-cs}x)ds, \quad B \in \mathcal{B}(\mathbb{R}^d), \tag{2.62}$$

$$\gamma = c^{-m}\gamma_\rho + c^{-m}\int_{|x|>1}\frac{x}{|x|}\sum_{j=0}^{m-1}\frac{(\log|x|)^j}{j!}\nu_\rho(dx), \tag{2.63}$$

and the correspondence of μ and ρ is one-to-one.

Proof By induction. When $m = 1$, the statements reduce to Theorem 2.17. Let $m \geq 2$. Assume that the assertions are true for $m-1$ in place of m. Let us show the assertions for m.

Step 1. We assume $\int_{|x|>2}(\log|x|)^m \nu_\rho(dx) < \infty$. Noting that $\rho \in ID_{\log}$, write $\mu_1 = \Phi_{(c)}(\rho)$, $\psi_1 = \log\widehat{\mu}_1$, and (A_1, ν_1, γ_1) for the triplet of μ_1. Then, using (2.30), we get

$$\begin{aligned}\int_{|x|>2}(\log|x|)^{m-1}\nu_1(dx) &= c^{-1}\int_{\mathbb{R}^d}\nu_\rho(dx)\int_0^\infty\left(\log|e^{-s}x|\right)^{m-1}1_{\{|e^{-s}x|>2\}}ds\\ &= c^{-1}\int_{|x|>2}\nu_\rho(dx)\int_0^{\log(|x|/2)}(\log|x|-s)^{m-1}ds\\ &= (cm)^{-1}\int_{|x|>2}\left[(\log|x|)^m-(\log 2)^m\right]\nu_\rho(dx)<\infty.\end{aligned} \tag{2.64}$$

It follows that $\mu_1 \in \mathfrak{D}(\Phi_{(c)}^{m-1})$. Thus $\rho \in \mathfrak{D}(\Phi_{(c)}^m)$. Let $\mu = \Phi_{(c)}^m(\rho)$. Since $\mu_1 \in L_0$, repeated application of Theorem 2.29 (i) shows that $\mu \in L_{m-1}$.

Let us show (2.59). Define η by

$$\eta(B) = \int_{\mathbb{R}^d}\nu_\rho(dx)\int_0^\infty s^{m-1}1_B(e^{-s}x)ds.$$

Then

$$\begin{aligned}\int_{|x|\le 1}|x|^2\eta(dx) &= \int_{\mathbb{R}^d}\nu_\rho(dx)\int_0^\infty s^{m-1}\left|e^{-s}x\right|^2 1_{\{|e^{-s}x|\le 1\}}ds\\ &= \int_{|x|\le 1}|x|^2\nu_\rho(dx)\int_0^\infty s^{m-1}e^{-2s}ds + \int_{|x|>1}|x|^2\nu_\rho(dx)\int_{\log|x|}^\infty s^{m-1}e^{-2s}ds,\end{aligned}$$

which is finite, since $\int_{\log|x|}^\infty s^{m-1}e^{-2s}ds \sim 2^{-1}|x|^{-2}(\log|x|)^{m-1}$ as $|x| \to \infty$. Moreover

$$\begin{aligned}\int_{|x|>1}\eta(dx) &= \int_{\mathbb{R}^d}\nu_\rho(dx)\int_0^\infty s^{m-1}1_{\{|e^{-s}x|>1\}}ds\\ &= m^{-1}\int_{|x|>1}(\log|x|)^m\nu_\rho(dx) < \infty.\end{aligned}$$

Hence, writing

$$\begin{aligned}\psi_\rho(e^{-cs}z) &= -\tfrac{1}{2}e^{-2cs}\langle z, A_\rho z\rangle + ie^{-cs}\langle\gamma_\rho, z\rangle\\ &\quad + \int_{\mathbb{R}^d}g(z, e^{-cs}x)\nu_\rho(dx) + i\int_{\mathbb{R}^d}\langle z, e^{-cs}x\rangle 1_{\{1<|x|\le e^{cs}\}}\nu_\rho(dx)\end{aligned} \tag{2.65}$$

with $g(z, x)$ of (1.19) and (2.39) and estimating in the same way as in the proof of Theorem 2.17 (i), we obtain (2.59). Here we also use the estimate

$$\int_0^\infty s^{m-1}e^{-s}ds\int_{1<|x|\le e^s}|x|\nu_\rho(dx)=\int_{|x|>1}|x|\nu_\rho(dx)\int_{\log|x|}^\infty s^{m-1}e^{-s}ds<\infty.$$

Since $\psi_1(z)=\int_0^\infty\psi_\rho(e^{-cs}z)ds$, we have

$$\begin{aligned}\psi(z)&=\int_0^\infty\frac{s^{m-2}}{(m-2)!}\psi_1(e^{-cs}z)ds=\int_0^\infty\frac{s^{m-2}}{(m-2)!}ds\int_s^\infty\psi_\rho(e^{-cu}z)du\\&=\int_0^\infty\psi_\rho(e^{-cu}z)du\int_0^u\frac{s^{m-2}}{(m-2)!}ds,\end{aligned}$$

which gives (2.60). Here we used Fubini's theorem, which is permitted by (2.59). Now, calculating the right-hand side of (2.60) from (2.65), we see that μ has triplet (A,ν,γ) described by (2.61)–(2.63). In obtaining (2.63) we also use

$$c^{-m}\int_{|x|>1}x\nu_\rho(dx)\int_{\log|x|}^\infty\frac{s^{m-1}}{(m-1)!}e^{-s}ds=c^{-m}\int_{|x|>1}\frac{x}{|x|}\sum_{j=0}^{m-1}\frac{(\log|x|)^j}{j!}\nu_\rho(dx).$$

Step 2. Suppose that we are given $\rho\in ID$ satisfying $\int_{|x|>2}(\log|x|)^m\nu_\rho(dx)=\infty$. We claim that $\rho\notin\mathfrak{D}(\Phi_{(c)}^m)$. From the definition we have

$$ID_{\log}=\mathfrak{D}(\Phi_{(c)})\supset\mathfrak{D}(\Phi_{(c)}^2)\supset\cdots.$$

Hence, if $\rho\notin ID_{\log}$, then $\rho\notin\mathfrak{D}(\Phi_{(c)}^m)$. So we may and do assume that $\rho\in ID_{\log}$. Let $\mu_1=\Phi_{(c)}(\rho)$ and ν_1 be the Lévy measure of μ_1. Then the same calculus as in (2.64) shows that $\int_{|x|>2}(\log|x|)^{m-1}\nu_1(dx)=\infty$. Hence $\mu_1\notin\mathfrak{D}(\Phi_{(c)}^{m-1})$ from the induction hypothesis. It follows that $\rho\notin\mathfrak{D}(\Phi_{(c)}^m)$. Combined with Step 1, this proves (2.57).

Step 3. Let μ be an arbitrary member of L_{m-1}. We claim that $\mu\in\mathfrak{R}(\Phi_{(c)}^m)$. Since $L_{m-1}\subset L_0=\mathfrak{R}(\Phi_{(c)})$, there is μ_{m-1} such that $\Phi_{(c)}(\mu_{m-1})=\mu$. Hence, by virtue of Theorem 2.29 (i), $\mu_{m-1}\in L_{m-2}\cap ID_{\log}$. Then $\mu_{m-1}\in\mathfrak{R}(\Phi_{(c)}^{m-1})$ from the induction hypothesis. Hence $\mu\in\mathfrak{R}(\Phi_{(c)}^m)$. Step 1 and this prove (2.58). We have the one-to-one property in the theorem since $\Phi_{(c)}$ is one-to-one. ■

Remark 2.33 The description (2.57) of $\mathfrak{D}(\Phi_{(c)}^m)$ can be written as

$$\mathfrak{D}(\Phi_{(c)}^m)=\left\{\rho\in ID\colon \int_{|x|>2}(\log|x|)^m\rho(dx)<\infty\right\}.\tag{2.66}$$

See [93] Theorem 25.3 and Proposition 25.4 for a proof.

The class L_{m-1} is directly expressed as the range of an improper stochastic integral mapping as follows.

Theorem 2.34 *Let $m \in \mathbb{N}$. Fix a constant $c_m > 0$. Let $f_m(s) = e^{-(c_m s)^{1/m}}$ and consider the mapping Φ_{f_m} in Definition 2.26. Then $\mathfrak{D}(\Phi_{f_m})$ and $\mathfrak{R}(\Phi_{f_m})$ do not depend on c_m. If $\Phi_{(c)}$ is with parameter c satisfying $c_m = m!\, c^m$, then*

$$\Phi_{f_m} = \Phi^m_{(c)}, \tag{2.67}$$

which means that $\mathfrak{D}(\Phi_{f_m}) = \mathfrak{D}(\Phi^m_{(c)})$ and that $\Phi_{f_m}(\rho) = \Phi^m_{(c)}(\rho)$ for $\rho \in \mathfrak{D}(\Phi_{f_m})$.

Proof As before, let $\psi_\rho(z) = (\log \widehat{\rho})(z)$. Then

$$\mathfrak{D}(\Phi_{f_m}) = \left\{\rho \in ID\colon \textstyle\int_0^\infty |\psi_\rho(f_m(s)z)| ds < \infty \text{ for } z \in \mathbb{R}^d\right\}. \tag{2.68}$$

This is shown in (2.52) for $m = 1$. For general $m \in \mathbb{N}$ Proposition 2.25 says (2.68) with "=" replaced by "$\supset$". The proof of (2.68) with "=" replaced by "$\subset$" is given in Theorem 6.3 (iii) of [101] or Proposition 4.3 of [99]. It follows from (2.68) via change of variables in integration that $\mathfrak{D}(\Phi_{f_m})$ does not depend on the choice of c_m. Now let us show $\mathfrak{D}(\Phi_{f_m}) = \mathfrak{D}(\Phi^m_{(c)})$. Choosing $c_m = m!\, c^m$, we have

$$\int_0^\infty |\psi_\rho(f_m(s)z)| ds = \int_0^\infty \frac{u^{m-1}}{(m-1)!} |\psi_\rho(e^{-cu}z)| du,$$

by change of variables $s = (m!)^{-1} u^m$. Hence $\mathfrak{D}(\Phi_{f_m}) \supset \mathfrak{D}(\Phi^m_{(c)})$ by Theorem 2.32 and (2.68). Conversely, suppose $\rho \in \mathfrak{D}(\Phi_{f_m})$. It follows from (2.68) that we can define

$$\psi_j(z) = \int_0^\infty \frac{u^{j-1}}{(j-1)!} \psi_\rho(e^{-cu}z) du, \quad j = 1, \ldots, m.$$

Then, for $j = 2, \ldots, m$, we see

$$\int_0^\infty |\psi_{j-1}(e^{-cs}z)| ds = \int_0^\infty ds \left| \int_0^\infty \frac{u^{j-2}}{(j-2)!} \psi_\rho(e^{-c(u+s)}z) du \right|$$

$$\leq \int_0^\infty ds \int_s^\infty \frac{(v-s)^{j-2}}{(j-2)!} |\psi_\rho(e^{-cv}z)| dv = \int_0^\infty \frac{v^{j-1}}{(j-1)!} |\psi_\rho(e^{-cv}z)| dv < \infty,$$

$$\int_0^\infty \psi_{j-1}(e^{-cs}z) ds = \int_0^\infty \frac{v^{j-1}}{(j-1)!} \psi_\rho(e^{-cv}z) dv = \psi_j(z).$$

It follows from Theorem 2.17 and Proposition 2.28 that there is μ_j satisfying $(\log \widehat{\mu}_j)(z) = \psi_j(z)$ and that $\mu_{j-1} \in \mathfrak{D}(\Phi_{(c)})$ and $\Phi_{(c)}(\mu_{j-1}) = \mu_j$ for $j = 1, \ldots, m$. Hence $\rho \in \mathfrak{D}(\Phi^m_{(c)})$ and $\Phi^m_{(c)}(\rho) = \Phi_{f_m}(\rho)$. ∎

Notes

The formula (2.30) of the Lévy measures of selfdecomposable distributions was found by Urbanik [127] (1969), although he did not recognize its probabilistic meaning. The representation of the class of selfdecomposable distributions as the class of limit distributions of OU type processes (Theorem 2.17) was discovered by Wolfe [139, 140] (1982a, 1982b) and, immediately after that, the papers Sato and Yamazato [109, 110] (1983, 1984), Jurek and Vervaat [45] (1983), and Gravereaux [28] (1982) were worked out. Wolfe's paper [139] (1982a) was submitted in October 1979. Sato and Yamazato used, in the paper [108] (1978) studying fine properties of the density of a general selfdecomposable distribution on $\mathbb{R}$, an integro-differential equation which showed that it was a stationary distribution density of an OU type process. Most of these results were obtained in the form of operator generalization, which will be touched upon in Sect. 5.2.

In the case where $\{Z_t\}$ is an increasing Lévy process on $\mathbb{R}$, the limit theorem in Theorem 2.17 was obtained earlier by Çinlar and Pinsky [20, 21] (1971, 1972) in storage theory. They treated a more general equation than (2.13), where cX_s is replaced by a release function $r(X_s) > 0$.

Theorem 2.29 was given by [45] (1983) and [109] (1983) for (i) on L_m and by [139] (1982a) and [45] (1983) for (ii) on $\mathfrak{S}$ and $\mathfrak{S}^0$. Theorem 2.32 on L_m was due to Jurek [40] (1983b) in a different formulation. The representation of L_m as the range of an improper stochastic integral in Theorem 2.34 is given in [40] (1983b). Another expression of $L_m(\mathbb{R})$ is made by Graversen and Pedersen [29] (2011).

Any OU type process on $\mathbb{R}^d$ having limit distribution (that is, satisfying (2.25)) is recurrent. But there are recurrent OU type processes without limit distribution. Also there are transient OU type processes. Indeed, suppose that $d = 1$ and that $\{X_t\}$ is an OU type process generated by $c > 0$ and ρ with Lévy measure $\nu(dx) = 1_{\{|x|>2\}}(x)b|x|^{-1}(\log|x|)^{-\alpha-1}dx$ with $\alpha > 0$, $b > 0$. Then (2.25) is satisfied if and only if $\alpha > 1$; if $\alpha < 1$, then $\{X_t\}$ is transient; if $\alpha = 1$ and $2b > c$, then $\{X_t\}$ is transient; if $\alpha = 1$ and $2b \leq c$, then $\{X_t\}$ is recurrent ([110] (1984)). A general criterion of recurrence and transience was given by Shiga [114] (1990) for $d = 1$ and by Sato, Watanabe, and Yamazato [105] (1994) for $d \geq 2$.

Study of the improper stochastic integral mapping Φ_f in Definition 2.26 was developed in Jurek [42] (1988), Maejima and Sato [60] (2003), Sato [96–101] (2004, 2006a, 2006b, 2006c, 2009, 2010), Barndorff-Nielsen, Maejima, and Sato [7] (2006), Barndorff-Nielsen, Rosiński, and Thorbjørnsen [10] (2008), Maejima [59] (2015), and others. One of the results is the continuous-parameter interpolation and extrapolation of the mappings Φ_{f_m}, $m \in \mathbb{N}$, with $f_m(s) = e^{-c(m!s)^{1/m}}$ in Theorem 2.34 to the mappings Φ_{f_p}, $p \in (0, \infty)$, with $f_p(s) = e^{-c(\Gamma(p+1)s)^{1/p}}$. This is connected with fractional integral of Riemann, Liouville, and others. It is true that $\Phi_{f_{p+q}} = \Phi_{f_q}\Phi_{f_p}$ for $p, q \in (0, \infty)$. Thu [123, 124] (1982, 1984) and [125] (1986) are pioneering works in this direction. The concept of function on $\mathbb{R}$ monotone of order $n \in \mathbb{N}$ in Definition 1.37 is extended to the concept of function monotone of order $p \in (0, \infty)$ and it is shown that $\mu \in \mathfrak{R}(\Phi_{f_p})$ if and only if $\mu \in L_0$ and its h-function $h_\xi(u)$ is monotone of order p in $u \in \mathbb{R}$. Another extension of the concept of selfdecomposability is to introduce the concept of function on $(0, \infty)$ monotone of order $n \in \mathbb{N}$ similarly to Definition 1.37, to extend it to monotone of order $p \in (0, \infty)$, and, for some function g_p, to show that $\mu \in \mathfrak{R}(\Phi_{g_p})$ if and only if $\mu \in L_0$ and its k-function $k_\xi(r)$ is monotone of order p in $r \in (0, \infty)$. The limiting class $\bigcap_p \mathfrak{R}(\Phi_{g_p})$ equals the Thorin class on $\mathbb{R}^d$, which will be explained below. The processes in Rosiński [83] (2007) are closely connected with distributions in the scheme of [101] (2010). The limit of the ranges of iterations of Φ_f is studied by Maejima and Sato [61] (2009) and Sato [102] (2011); they are extension of the fact $L_\infty = \bigcap_{m=1}^{\infty} \mathfrak{R}(\Phi_{(c)}^m)$.

After the study of the roles of Gaussian, stable, infinitely divisible, selfdecomposable, and compound Poisson distributions in the early stage of modern probability theory, new sufficient conditions for infinite divisibility, new explicit distributions in ID appearing in stochastic processes, and new subclasses of ID were obtained. It was from 1960s on by Goldie [27] (1967), Steutel [116, 117] (1967, 1970), Thorin [120, 121] (1977a, 1977b), Bondesson [15] (1981), Barndorff-Nielsen and Halgreen (1977) [6], Barndorff-Nielsen [4] (1978), Halgreen [30] (1979), and others. For one-dimensional distributions Bondesson [16] (1992) and Steutel and van Harn [118] (2004) are monographs with detailed information. Some of the subclasses can be expressed as $\mathfrak{R}(\Phi_f)$. We mention two important classes: Thorin class $T = T(\mathbb{R}^d)$ and Goldie-Steutel-Bondesson class $B = B(\mathbb{R}^d)$. Class T is the $\mathbb{R}^d$-version of the class GGC (generalized Γ-convolutions) introduced by Thorin in the course of his proof of infinite divisibility of Pareto and log-normal distributions. Class B is the $\mathbb{R}^d$-version of the class *gcmed* (generalized convolutions of mixture of exponential distributions) studied by Bondesson. A distribution $\mu \in \mathfrak{P}(\mathbb{R}^d)$ is in T if and only if $\widehat{\mu}(z)$ is of the form (1.25) with $k_\xi(r)$ being completely monotone in $r \in (0, \infty)$; μ is in B if and only if its Lévy measure is of the form (1.27) with ν_ξ having completely monotone density in $(0, \infty)$ for each ξ. Let us define a mapping Ψ by $\Psi = \Phi_f$ with $f(s)$ satisfying $s = \int_{f(s)}^{\infty} u^{-1} e^{-u} du$. Then we can prove $\mathfrak{D}(\Psi) = ID_{\log}$. Further, let us define a different type of *improper stochastic integral* $\int_{0+}^{t_0} f(s) dZ_s$ with a finite t_0 as the limit in probability of $\int_t^{t_0} f(s) dZ_s$ as $t \to 0+$. The mapping $\Upsilon(\rho) = \mathcal{L}\left(\int_{0+}^{1} \log(1/s) dZ_s^{(\rho)}\right)$ was introduced by Barndorff-Nielsen and Thorbjørnsen [11] (2004) in an area called free probability. It is easy to show that $\mathfrak{D}(\Upsilon) = ID$. Barndorff-Nielsen, Maejima, and Sato [7] (2006) showed that $\Phi_{(1)}\Upsilon = \Upsilon\Phi_{(1)} = \Psi$, $B = \mathfrak{R}(\Upsilon)$, and $T = \Phi_{(1)}(B) = \Upsilon(L_0) = \mathfrak{R}(\Psi)$. Several works followed this result. James, Roynette, and Yor [37] (2008) contains another characterization and examples of class T.

Alf and O'Connor [2] (1977) showed that $\mu \in ID(\mathbb{R})$ has Lévy measure unimodal with mode 0 if and only if $(\log \widehat{\mu})(z) = z^{-1} \int_0^z (\log \widehat{\rho})(u) du$ for some $\rho \in ID$, which is equivalent to $\mu = \mathcal{L}\left(\int_0^1 s dZ_s^{(\rho)}\right)$. O'Connor [67, 68] (1979a, 1979b) called such μ as of class U and studied some classes with a continuous parameter which satisfy a relation similar to selfdecomposability (1.3). Jurek [41–43] (1985, 1988, 1989) obtained similar results and proved their stochastic integral representation. They are contained in the scheme of [101] (2010) and the $\mathbb{R}^d$-version of O'Connor's U is called *Jurek class* $U = U(\mathbb{R}^d)$. Further, Maejima and Ueda [65] (2010) found a temporally inhomogeneous extension of OU type processes related to those continuous-parameter classes.

Chapter 3
Selfsimilar Additive Processes and Stationary Ornstein–Uhlenbeck Type Processes

Selfsimilar processes on $\mathbb{R}^d$ are those stochastic processes whose finite-dimensional distributions are such that any change of time scale has the same effect as some change of spatial scale. Under the condition of stochastic continuity and non-zero, the relation of the two scale changes is expressed by a positive number c called exponent. A Lévy process is selfsimilar if and only if it is strictly stable.

In Sect. 3.1 selfsimilar additive processes are studied. It is shown that the class of those processes on $\mathbb{R}^d$ exactly corresponds to the class $L_0(\mathbb{R}^d)$. By a transformation named after Lamperti, selfsimilar processes correspond to stationary processes. In the case of selfsimilar additive processes this is realized by the correspondence to stationary OU type processes. Those results will be given in Sect. 3.2.

3.1 Selfsimilar Additive Processes and Class L_0

It will be shown that, for any selfsimilar additive process $\{X_t\}$ on $\mathbb{R}^d$, the distribution of X_t is selfdecomposable for every t, that is, $\mathcal{L}(X_t) \in L_0\left(\mathbb{R}^d\right)$. Conversely, if μ belongs to the class $L_0\left(\mathbb{R}^d\right)$, then for every $c > 0$, there is, uniquely in law, a c-selfsimilar additive process $\{X_t\}$ on $\mathbb{R}^d$ with $\mathcal{L}(X_1) = \mu$. As a consequence, there are many additive processes which are selfsimilar.

Definition 3.1 A stochastic process $\{X_t : t \geq 0\}$ on $\mathbb{R}^d$ is *selfsimilar* if, for any $a > 0$, there is $b > 0$ such that $\{X_{at} : t \geq 0\} \stackrel{d}{=} \{bX_t : t \geq 0\}$.

Theorem 3.2 *Let $\{X_t : t \geq 0\}$ be a selfsimilar, stochastically continuous, non-zero process on $\mathbb{R}^d$ with $X_0 = 0$ a.s. Then b in the definition above is uniquely determined by a and there is $c > 0$ such that, for any $a > 0$, $b = a^c$.*

See Theorem 13.11 and Remark 13.13 of [93].

A. Rocha-Arteaga, K. Sato, *Topics in Infinitely Divisible Distributions and Lévy Processes*, SpringerBriefs in Probability and Mathematical Statistics,
https://doi.org/10.1007/978-3-030-22700-5_3

Definition 3.3 A stochastic process $\{X_t : t \geq 0\}$ on $\mathbb{R}^d$ is called a *c-selfsimilar* process if it is stochastically continuous, starts from 0, and satisfies $\{X_{at} : t \geq 0\} \stackrel{d}{=} \{a^c X_t : t \geq 0\}$ for all $a > 0$. The number $c > 0$ is called the *exponent of selfsimilarity*. (If in addition $\{X_t\}$ is non-zero, then its exponent c is uniquely determined. But, if $\{X_t\}$ is a zero process, it is c-selfsimilar for all $c > 0$.)

We study selfsimilar additive processes on $\mathbb{R}^d$. Recall that if $\{X_t : t \geq 0\}$ is an additive process, then $\mathcal{L}(X_t) \in ID$ for any $t \geq 0$ (Theorem 1.32). First, let us consider time change by powers of t.

Proposition 3.4 *If $\{X_t\}$ is a c-selfsimilar additive process on $\mathbb{R}^d$, then, for any $\kappa > 0$, $\{X_{t^\kappa}\}$ is a $c\kappa$-selfsimilar additive process.*

Proof We can prove that $\{X_{t^\kappa}\}$ is an additive process from the assumption that $\{X_t\}$ is additive. If $\{X_t\}$ is c-selfsimilar, then $\{X_{(at)^\kappa}\} = \{X_{a^\kappa t^\kappa}\} \stackrel{d}{=} \{a^{c\kappa} X_{t^\kappa}\}$ and thus $\{X_{t^\kappa}\}$ is $c\kappa$-selfsimilar. ■

This shows that exponents are not important for the study of properties of selfsimilar additive processes because we can freely change the exponent c by a simple time change.

The following theorem establishes the relation between selfsimilar additive processes and selfdecomposable distributions.

Theorem 3.5

(i) *Let $\{X_t : t \geq 0\}$ be a selfsimilar additive process on $\mathbb{R}^d$. Then $\mathcal{L}(X_t) \in L_0(\mathbb{R}^d)$ for all $t \geq 0$.*
(ii) *For any $\mu \in L_0(\mathbb{R}^d)$ and any $c > 0$, there is a unique (in law) c-selfsimilar additive process $\{X_t\}$ on $\mathbb{R}^d$ such that $\mathcal{L}(X_1) = \mu$.*

Proof

(i) Let c be the exponent of selfsimilarity of $\{X_t\}$. For $0 \leq s \leq t$ let μ_t and $\mu_{s,t}$ be the distributions of X_t and $X_t - X_s$, respectively. We have

$$\widehat{\mu}_t(z) = \widehat{\mu}_s(z)\,\widehat{\mu}_{s,t}(z) = \widehat{\mu}_t\left((s/t)^c z\right)\widehat{\mu}_{s,t}(z), \tag{3.1}$$

by the independent increments and by $X_s = X_{(s/t)t} \stackrel{d}{=} (s/t)^c X_t$. Given $b > 1$ choose $0 < s < t$ such that $b = (s/t)^{-c}$. Then the identity above shows that $\mu_t \in L_0$ for $t > 0$. Since $\mathcal{L}(X_0) = \delta_0$, $\mu_0 \in L_0$ is evident.
(ii) Suppose that $\mu \in L_0$ and $c > 0$ are given. Then $\widehat{\mu}(z) \neq 0$ and, for every $b > 1$, there is a unique $\eta_b \in ID$ such that $\widehat{\mu}(z) = \widehat{\mu}\left(b^{-1}z\right)\widehat{\eta}_b(z)$. Next define μ_t and $\mu_{s,t}$ by $\mu_0 = \delta_0$,

$$\widehat{\mu}_t(z) = \widehat{\mu}\left(t^c z\right) \qquad \text{for } t > 0,$$

$$\widehat{\mu}_{s,t}(z) = \widehat{\eta}_{(t/s)^c}\left(t^c z\right) \qquad \text{for } 0 < s < t$$

and $\mu_{0,t} = \mu_t$ for $t > 0$. We have $\mu_t = \mu_s * \mu_{s,t}$ for $0 \le s < t$, since

$$\widehat{\mu}\left(t^c z\right) = \widehat{\mu}\left(s^c z\right) \widehat{\eta}_{(t/s)^c}\left(t^c z\right) \qquad \text{for } 0 < s < t.$$

Therefore $\widehat{\mu}_{s,t}(z) = \widehat{\mu}_t(z)/\widehat{\mu}_s(z)$. It follows that, for $0 \le r < s < t$,

$$\widehat{\mu}_{r,s}(z)\,\widehat{\mu}_{s,t}(z) = \frac{\widehat{\mu}_s(z)\,\widehat{\mu}_t(z)}{\widehat{\mu}_r(z)\,\widehat{\mu}_s(z)} = \frac{\widehat{\mu}_t(z)}{\widehat{\mu}_r(z)} = \widehat{\mu}_{r,t}(z),$$

that is, $\mu_{r,s} * \mu_{s,t} = \mu_{r,t}$. Now Kolmogorov's extension theorem applies and we can construct, on some probability space, a process $\{X_t : t \ge 0\}$ such that, for $0 \le t_0 < t_1 < \cdots < t_n$ and $B_0, \ldots, B_n \in \mathcal{B}(\mathbb{R}^d)$,

$$P\left[X_{t_0} \in B_0, \ldots, X_{t_n} \in B_n\right] = \int \mu_{t_0}(dx_0) 1_{B_0}(x_0) \int \mu_{t_0,t_1}(dx_1)$$

$$1_{B_1}(x_0 + x_1) \int \cdots \int \mu_{t_{n-1},t_n}(dx_n) 1_{B_n}(x_0 + \cdots + x_{n-1} + x_n).$$

We have $\mathcal{L}(X_t) = \mu_t$. In particular, $\mathcal{L}(X_1) = \mu$. The process $\{X_t\}$ starts at 0 a. s., has independent increments, and is stochastically continuous because $\mu_{s,t} \to \delta_0$ as $s \uparrow t$ or $t \downarrow s$. Indeed, $\widehat{\eta}_b(z) \to 1$ uniformly in any neighbourhood of 0 as $b \downarrow 1$. Therefore it is an additive process in law. By choosing a modification, it is an additive process ([93] Theorem 11.5). Moreover, we have from the definition of μ_t that

$$X_{ut} \stackrel{d}{=} a^c X_t \qquad \text{for } t \ge 0.$$

This implies that $\{X_{at}\}$ and $\{a^c X_t\}$ have a common system of finite-dimensional distributions, since both are additive processes. Here we have used Theorem 1.32 (ii). Hence $\{X_t\}$ is c-selfsimilar. Since $X_t \stackrel{d}{=} t^c X_1$, the process $\{X_t\}$ with the properties required is unique in law, again by Theorem 1.32 (ii). ∎

Proposition 3.6

(i) Let $0 < \alpha \le 2$. Suppose that $\{X_t\}$ is a strictly α-stable process on $\mathbb{R}^d$. Then $\{X_t\}$ is $(1/\alpha)$-selfsimilar.

(ii) Let $c > 0$. Suppose that $\{X_t\}$ is a non-zero c-selfsimilar Lévy process on $\mathbb{R}^d$. Then $c \ge 1/2$ and $\{X_t\}$ is a strictly $(1/c)$-stable process.

Proof

(i) Let $\mu = \mathcal{L}(X_1)$. Then $\mathcal{L}(X_t) = \mu^t \in \mathfrak{S}_\alpha^0$, that is, $\widehat{\mu}^{ta}(z) = \widehat{\mu}^t(a^{1/\alpha} z)$. Hence $X_{at} \stackrel{d}{=} a^{1/\alpha} X_t$. Since $\{X_{at}\}$ and $\{a^{1/\alpha} X_t\}$ are both Lévy processes, it follows that $\{X_{at}\} \stackrel{d}{=} \{a^{1/\alpha} X_t\}$.

(ii) We have $X_{at} \stackrel{d}{=} a^c X_t$ for all $t \geq 0$. In particular $\mu = \mathcal{L}(X_1)$ satisfies $\widehat{\mu}^a(z) = \widehat{\mu}(a^c z)$. Since $\mu \neq \delta_0$, Proposition 1.22 implies that $c = 1/\alpha$ with $0 < \alpha \leq 2$. Thus $\mu \in \mathfrak{S}_\alpha^0$. Hence $\{X_t\}$ is a strictly α-stable process. ■

Notice that, in the case of a selfsimilar Lévy process, the exponent of selfsimilarity is important. If $\{X_t\}$ is a non-zero Lévy process and $\kappa \neq 1$, then the time change in Proposition 3.4 transforms $\{X_t\}$ to a non-Lévy additive process.

Example 3.7 Let $0 < \alpha \leq 2$ and $\mu \in \mathfrak{S}_\alpha(\mathbb{R}^d)$. Let $\{X_t : t \geq 0\}$ be the $(1/\alpha)$-selfsimilar additive process with $\mathcal{L}(X_1) = \mu$ in Theorem 3.5. Let $\{Y_t : t \geq 0\}$ be the Lévy process with $\mathcal{L}(Y_1) = \mu$.

(i) Suppose that $\mu \in \mathfrak{S}_\alpha^0$. Then $\{X_t\} \stackrel{d}{=} \{Y_t\}$,
(ii) Suppose that $\mu \in \mathfrak{S}_\alpha \setminus \mathfrak{S}_\alpha^0$. Then $\{X_t\}$ and $\{Y_t\}$ are not identical in law and they are related as follows. If $\alpha \neq 1$, then

$$\{X_t\} \stackrel{d}{=} \left\{ Y_t + \left(t^{1/\alpha} - t\right)\tau \right\}, \tag{3.2}$$

where $\tau \neq 0$. Here τ is the drift of μ for $0 < \alpha < 1$ and the mean of μ for $1 < \alpha \leq 2$. If $\alpha = 1$, then

$$\{X_t\} \stackrel{d}{=} \left\{ Y_t - (t \log t)\, c_1 \int_S \xi \lambda\,(d\xi) \right\}, \tag{3.3}$$

where c_1 and λ denote c and λ in the expression (1.40) of the Lévy measure ν of μ, and we have $\int_S \xi \lambda(d\xi) \neq 0$.

Here is a proof. The identity in law in (i) and the first assertion in (ii) follow from Proposition 3.6. Let us prove the remaining part, using Theorem 1.42. Let $\mu_t = \mathcal{L}(X_t)$. Then $\widehat{\mu}_t(z) = \widehat{\mu}(t^{1/\alpha} z)$. We have $\mathcal{L}(Y_t) = \mu^t$.

Let $\alpha \neq 1$. Then $\widehat{\mu}(z) = \widehat{\eta}(z) e^{i\langle \tau, z\rangle}$, $\tau \neq 0$, and $\eta \in \mathfrak{S}_\alpha^0$. Hence

$$\widehat{\mu}_t(z) = \widehat{\eta}(t^{1/\alpha} z) e^{it^{1/\alpha}\langle \tau, z\rangle} = \widehat{\eta}(z)^t e^{it^{1/\alpha}\langle \tau, z\rangle} = \widehat{\mu}(z)^t e^{i(t^{1/\alpha} - t)\langle \tau, z\rangle},$$

that is, $X_t \stackrel{d}{=} Y_t + (t^{1/\alpha} - t)\tau$.

Let $\alpha = 1$. Then,

$$\widehat{\mu}(z) = \exp\left[c_1 \int_S \lambda(d\xi) \int_0^\infty \left(e^{ir\langle \xi, z\rangle} - 1 - ir\langle \xi, z\rangle 1_{(0,1]}(r) \right) \frac{dr}{r^2} + i\langle \gamma, z\rangle \right]$$

with $\int_S \xi \lambda(d\xi) \neq 0$. Since $\widehat{\mu}_t(z) = \widehat{\mu}(tz)$, we have

$$\widehat{\mu}_t(z) = \exp\left[c_1 \int_S \lambda(d\xi) \int_0^\infty \left(e^{iu\langle \xi, z\rangle} - 1 - iu\langle \xi, z\rangle 1_{(0,1]}\left(\tfrac{u}{t}\right) \right) \frac{t\,du}{u^2} + it\langle \gamma, z\rangle \right].$$

Noting that

$$1_{(0,1]}\left(\frac{u}{t}\right)=\begin{cases}1_{(0,1]}(u)-1_{(t,1]}(u) & \text{for } t<1\\ 1_{(0,1]}(u)+1_{(1,t]}(u) & \text{for } t>1,\end{cases}$$

we obtain

$$\widehat{\mu}_t(z)=\widehat{\mu}(z)^t \exp\left[-i(t\log t)\left\langle c_1\int_S \xi\lambda(d\xi), z\right\rangle\right],$$

that is, $X_t \stackrel{d}{=} Y_t-(t\log t)c_1\int_S \xi\lambda(d\xi)$. Now we get (3.2) for $\alpha\neq 1$ and (3.3) for $\alpha=1$, using Theorem 1.32 (ii), since both sides are additive processes.

When $\{X_t\}$ is a selfsimilar additive process, the distribution of X_t is selfdecomposable. But its joint distributions (finite-dimensional distributions) are not always selfdecomposable. Let us give conditions for joint distributions of $\{X_t\}$ to be selfdecomposable and, furthermore, conditions for them to be of class L_m.

Theorem 3.8 *Let $\mu\in L_0(\mathbb{R}^d)$ such that $\mu\neq\delta_0$. Let $c>0$ and let $\{X_t\}$ be the c-selfsimilar additive process with $\mathcal{L}(X_1)=\mu$. Let $m\in\{0,1,\dots,\infty\}$. Then the following six conditions are equivalent. Here we understand $L_{-1}=ID$ and $L_{\infty-1}=L_\infty$.*

(a) $\mu\in L_m(\mathbb{R}^d)$.
(b) $\mathcal{L}(X_t)\in L_m(\mathbb{R}^d)$ *for all* $t\geq 0$.
(c) $\mathcal{L}\left((X_{t_k})_{k=1,2}\right)\in L_{m-1}(\mathbb{R}^{2d})$ *for all* $t_1,t_2\geq 0$.
(d) $\mathcal{L}(c_1X_{t_1}+c_2X_{t_2})\in L_{m-1}(\mathbb{R}^d)$ *for all* $t_1,t_2\geq 0$ *and* $c_1,c_2\in\mathbb{R}$.
(e) $\mathcal{L}\left((X_{t_k})_{1\leq k\leq n}\right)\in L_{m-1}(\mathbb{R}^{nd})$ *for all* $n\in\mathbb{N}$ *and* $t_1,\dots,t_n\geq 0$.
(f) $\mathcal{L}(c_1X_{t_1}+\cdots+c_nX_{t_n})\in L_{m-1}(\mathbb{R}^d)$ *for all* $n\in\mathbb{N}$, $t_1,\dots,t_n\geq 0$, *and* $c_1,\dots,c_n\in\mathbb{R}$.

Let us prepare

Lemma 3.9 *Let* $m\in\{0,1,\dots,\infty\}$.

(i) Let X be a random variable on $\mathbb{R}^{d_1}$ and F a real $d_2\times d_1$ matrix. If $\mathcal{L}(X)\in L_m(\mathbb{R}^{d_1})$, then $\mathcal{L}(FX)\in L_m(\mathbb{R}^{d_2})$.
(ii) Let $d_1,\dots,d_n\in\mathbb{N}$ and $d=d_1+\cdots+d_n$. Let X_j be a random variable on $\mathbb{R}^{d_j}$ for each j. If $X_1,\dots,X_n$ are independent and if $\mathcal{L}(X_j)\in L_m(\mathbb{R}^{d_j})$ for each j, then $\mathcal{L}((X_j)_{1\leq j\leq n})\in L_m(\mathbb{R}^d)$.

Proof

(i) Let $Y=FX$, $\mu=\mathcal{L}(X)$, and $\mu_Y=\mathcal{L}(Y)$. Suppose that $\mu\in L_0$. For any $b>1$, $\widehat{\mu}(z)=\widehat{\mu}(b^{-1}z)\widehat{\eta}_b(z)$ with some $\eta_b\in ID$. Let $F^\top$ be the transpose of F. Since $\widehat{\mu}_Y(z)=\widehat{\mu}(F^\top z)$, we have $\widehat{\mu}_Y(z)=\widehat{\mu}_Y(b^{-1}z)\widehat{\eta}_b(F^\top z)$. Hence $\mu_Y\in L_0$. This proves the assertion for $m=0$. By induction we can prove it for $m=1,2\dots$. The validity for $m=\infty$ follows from this.

(ii) Let $X_1, \ldots, X_n$ be independent, $X = (X_j)_{1\le j\le n}$, $\mu_j = \mathcal{L}(X_j)$, and $\mu = \mathcal{L}(X)$. Assume $\mu_j \in L_0$ for each j. Then $\widehat{\mu}_j(z_j) = \widehat{\mu}_j(b^{-1}z_j)\widehat{\eta}_{j,b}(z_j)$ for $z_j \in \mathbb{R}^{d_j}$ with some $\eta_{j,b} \in ID$. For $z = (z_j)_{1\le j\le d}$ with $z_j \in \mathbb{R}^{d_j}$,

$$\widehat{\mu}(z) = \prod_{j=1}^{n} \widehat{\mu}_j(z_j) = \prod_{j=1}^{n} \widehat{\mu}_j(b^{-1}z_j)\widehat{\eta}_{j,b}(z_j) = \widehat{\mu}(b^{-1}z)\widehat{\eta}_b(z),$$

where $\widehat{\eta}_b(z) = \prod_{j=1}^{n} \widehat{\eta}_{j,b}(z_j)$. Hence the assertion is true for $m = 0$. It is true for $m = 1, 2 \ldots$ by induction. Thus it is true also for $m = \infty$. ■

Proof of Theorem 3.8 Conditions (a) and (b) are equivalent, because $\mathcal{L}(X_t) = \mathcal{L}(t^c X_1)$ for $t > 0$ and $\mathcal{L}(X_0) = \delta_0$.

Let $0 \le s < t$ and let $\mu_t = \mathcal{L}(X_t)$ and $\mu_{s,t} = \mathcal{L}(X_t - X_s)$. Then (3.1) shows that $\mu_t \in L_m$ if and only if $\mu_{s,t} \in L_{m-1}$ for $0 < s < t$.

Now let us prove that (b)⇒(e)⇒(f)⇒(d)⇒(b) and that (e)⇒(c)⇒(d).

(b)⇒(e): We have $\mathcal{L}(X_t - X_s) \in L_{m-1}$ for all $0 \le s \le t$. Given $0 = t_0 \le t_1 \le \cdots \le t_n$, we see that $X_{t_k} - X_{t_{k-1}}$, $k = 1, \ldots, n$, are independent and hence, by Lemma 3.9 (ii), $\mathcal{L}\left((X_{t_k} - X_{t_{k-1}})_{1\le k\le n}\right) \in L_{m-1}$. Since there is some $n \times n$ matrix F satisfying $(X_{t_k})_{1\le k\le n} = F(X_{t_k} - X_{t_{k-1}})_{1\le k\le n}$, we see $\mathcal{L}\left((X_{t_k})_{1\le k\le n}\right) \in L_{m-1}$ by Lemma 3.9 (i).

(e)⇒(f): This follows from Lemma 3.9 (i), since $c_1 X_{t_1} + \cdots + c_n X_{t_n} = F(X_{t_k})_{1\le k\le n}$ with some $1 \times n$ matrix F.

(f)⇒(d): (d) is a special case of (f).

(d)⇒(b): Since $X_t - X_s$ is a special case of $c_1 X_{t_1} + c_2 X_{t_2}$, we have $\mu_{s,t} \in L_{m-1}$ for $0 < s < t$.

(e)⇒(c): (c) is a special case of (e).

(c)⇒(d): Use Lemma 3.9 (i) as in the proof that (e)⇒(f). ■

Remark 3.10 The result in Theorem 3.8 on selfsimilar additive processes is quite different from Lévy processes. Let $m \in \{0, 1, \ldots, \infty\}$. If $\{Y_t\}$ is a Lévy process on $\mathbb{R}^d$ with $\mathcal{L}(Y_1) \in L_m(\mathbb{R}^d)$, then $\mathcal{L}((Y_{t_k})_{1\le k\le n}) \in L_m(\mathbb{R}^{nd})$ for every $n \in \mathbb{N}$ and $t_1, \ldots, t_n$ in $[0, \infty)$. This follows from Proposition 1.18 (iv) and Lemma 3.9.

3.2 Lamperti Transformation and Stationary Ornstein–Uhlenbeck Type Processes

We define stationary processes as follows. Notice that the set of time parameter is the whole line $\mathbb{R}$.

Definition 3.11 A stochastic process $\{Y_s : s \in \mathbb{R}\}$ on $\mathbb{R}^d$ is called *stationary process* if

$$\{Y_s : s \in \mathbb{R}\} \stackrel{d}{=} \{Y_{s+u} : s \in \mathbb{R}\} \quad \text{for } u \in \mathbb{R},$$

that is, if shift of time preserves the system of finite-dimensional distributions.

We use the following simple fact.

Lemma 3.12

(i) Let $(X_j)_{1\le j\le N}$ and $(X'_j)_{1\le j\le N}$ be random variables on $\mathbb{R}^N$ with a common law. Let Q be an $N' \times N$ real matrix. Then $Q(X_j)_{1\le j\le N} \stackrel{d}{=} Q(X'_j)_{1\le j\le N}$.

(ii) Let $\{X_j : 1 \le j \le N\}$ and $\{X'_j : 1 \le j \le N\}$ be families of random variables on $\mathbb{R}^d$ such that $(X_j)_{1\le j\le N} \stackrel{d}{=} (X'_j)_{1\le j\le N}$. Then, for any $a_1, \ldots, a_N \in \mathbb{R}$, we have $(a_j X_j)_{1\le j\le N} \stackrel{d}{=} (a_j X'_j)_{1\le j\le N}$.

Proof

(i) Since $\langle z, Qx\rangle = \langle Q^\top z, x\rangle$ for $z \in \mathbb{R}^{N'}$ and $x \in \mathbb{R}^N$, we have

$$Ee^{i\langle z, Q(X_j)_{1\le j\le N}\rangle} = Ee^{i\langle Q^\top z, (X_j)_{1\le j\le N}\rangle} = Ee^{i\langle Q^\top z, (X'_j)_{1\le j\le N}\rangle}$$
$$= Ee^{i\langle z, Q(X'_j)_{1\le j\le N}\rangle}.$$

(ii) We can consider $\{X_j : 1 \le j \le N\}$ and $\{X'_j : 1 \le j \le N\}$ as random variables on $\mathbb{R}^{Nd}$. Choosing a suitable $Nd \times Nd$ matrix Q, we see that the assertion is a special case of (i). ■

In order to make the situation clear, let us introduce some classes of stochastic processes. We identify two stochastic processes which are identical in law. Given $c > 0$, let $\mathbf{Ss}_c = \mathbf{Ss}_c(\mathbb{R}^d)$ be the class of c-selfsimilar processes on $\mathbb{R}^d$, which are defined in Definition 3.3. Let $\mathbf{St} = \mathbf{St}(\mathbb{R}^d)$ be the class of stochastically continuous stationary processes on $\mathbb{R}^d$. Unlike a selfsimilar process, a general process in $\mathbf{St}$ does not have any "exponent".

Theorem 3.13

(i) Let $c > 0$ and let $\{X_t : t \ge 0\} \in \mathbf{Ss}_c$. Let $\{Y_s : s \in \mathbb{R}\}$ be the process defined by

$$Y_s = e^{-cs} X_{\exp s} \quad \textit{for} \quad s \in \mathbb{R}. \tag{3.4}$$

Then $\{Y_s : s \in \mathbb{R}\} \in \mathbf{St}$.

(ii) Let $\{Y_s : s \in \mathbb{R}\} \in \mathbf{St}$. Let $c > 0$ and let $\{X_t : t \ge 0\}$ be defined by

$$X_0 = 0 \quad \textit{and} \quad X_t = t^c Y_{\log t} \quad \textit{for} \quad t > 0. \tag{3.5}$$

Then $\{X_t : t \ge 0\} \in \mathbf{Ss}_c$.

Proof

(i) We have $\{X_{at}: t \geq 0\} \stackrel{d}{=} \{a^c X_t: t \geq 0\}$ for all $a > 0$. Hence $\{X_t: t \geq 0\} = \{X_{aa^{-1}t}: t \geq 0\} \stackrel{d}{=} \{a^c X_{a^{-1}t}: t \geq 0\}$. Hence $\{X_t: t \geq 0\} \stackrel{d}{=} \{e^{-cu} X_{(\exp u)t}: t \geq 0\}$ for all $u \in \mathbb{R}$. Using Lemma 3.12, we obtain

$$\begin{aligned}\{Y_s: s \in \mathbb{R}\} &= \{e^{-cs} X_{\exp s}: s \in \mathbb{R}\} \stackrel{d}{=} \{e^{-cs} e^{-cu} X_{(\exp u)(\exp s)}: s \in \mathbb{R}\} \\ &= \{Y_{s+u}: s \in \mathbb{R}\} \quad \text{for all } u,\end{aligned}$$

which shows that $\{Y_s\}$ is a stationary process. Its stochastic continuity follows from that of $\{X_t\}$.

(ii) We have $\{Y_s: s \in \mathbb{R}\} \stackrel{d}{=} \{Y_{s+u}: s \in \mathbb{R}\}$ for all u. Then, choosing $u = \log a$ and using Lemma 3.12, we have

$$\{X_{at}: t > 0\} = \{a^c t^c Y_{\log(at)}: t > 0\} \stackrel{d}{=} \{a^c t^c Y_{\log t}: t > 0\} = \{a^c X_t: t > 0\}$$

for all $a > 0$. The stochastic continuity of $\{X_t\}$ follows, for $t > 0$, from that of $\{Y_s\}$ and, for $t = 0$, from $X_t = t^c Y_{\log t} \stackrel{d}{=} t^c Y_0 \to 0$ in probability as $t \downarrow 0$. ■

Definition 3.14 Let Λ_c denote the mapping of $\{X_t: t \geq 0\}$ to $\{Y_s: s \in \mathbb{R}\}$ in Theorem 3.13 (i) and let Λ'_c denote the mapping of $\{Y_s: s \in \mathbb{R}\}$ to $\{X_t: t \geq 0\}$ in Theorem 3.13 (ii). Both Λ_c and Λ'_c are called *Lamperti transformation*.

Remark 3.15 It is straightforward to see that $\Lambda'_c \circ \Lambda_c$ is the identity mapping of $\mathbf{Ss}_c$ and that $\Lambda_c \circ \Lambda'_c$ is the identity mapping of $\mathbf{St}$, where the symbol $\circ$ denotes composition of mappings. Hence, Λ_c is a one-to-one mapping from $\mathbf{Ss}_c$ onto $\mathbf{St}$, Λ'_c is a one-to-one mapping from $\mathbf{St}$ onto $\mathbf{Ss}_c$, and

$$\Lambda'_c = (\Lambda_c)^{-1}, \qquad \Lambda_c = \left(\Lambda'_c\right)^{-1}.$$

Notice that the domain of Λ_c depends on c but its range does not depend on c; on the other hand, the domain of Λ'_c does not depend on c but its range depends on c.

There are two mappings from $\mathbf{Ss}_c$ into $\mathbf{Ss}_{c\kappa}$.

Proposition 3.16 *Let $c > 0$ and $\kappa > 0$. Let $\{X_t: t \geq 0\}$ be a c-selfsimilar process on $\mathbb{R}^d$. Define X'_t by $X'_t = X_{t^\kappa}$ for $t \geq 0$ and X''_t by $X''_t = t^{c\kappa - c} X_t$ for $t > 0$ and $X''_0 = 0$. Then both $\{X'_t: t \geq 0\}$ and $\{X''_t: t \geq 0\}$ are $c\kappa$-selfsimilar processes.*

Proof The assertion on $\{X'_t\}$ is proved in the same way as the proof of Proposition 3.4. For the process $\{X''_t\}$ we have

$$\{X''_{at}\} = \{(at)^{c\kappa - c} X_{at}\} \stackrel{d}{=} \{(at)^{c\kappa - c} a^c X_t\} = \{a^{c\kappa} t^{c\kappa - c} X_t\} = \{a^{c\kappa} X''_t\}$$

for $a > 0$ by Lemma 3.12. Clearly, $\{X''_t\}$ is stochastically continuous for $t > 0$. Since $X''_t \stackrel{d}{=} t^{c\kappa} X''_1 \to 0$ in probability as $t \downarrow 0$, $\{X''_t\}$ is stochastically continuous also at $t = 0$. ■

Definition 3.17 Let $I_c^{(\kappa)}$ denote the mapping of $\{X_t\}$ to $\{X'_t\}$ and $J_c^{(\kappa)}$ the mapping of $\{X_t\}$ to $\{X''_t\}$ in Proposition 3.16. Let $F^{(\kappa)}$ with $\kappa > 0$ denote the mapping of $\{Y_s : s \in \mathbb{R}\} \in \mathbf{St}$ to $\{Y_{\kappa s} : s \in \mathbb{R}\} \in \mathbf{St}$.

Proposition 3.4 shows that, if $\{X_t\}$ is a c-selfsimilar additive process, then $I_c^{(\kappa)}(\{X_t\})$ is a $c\kappa$-selfsimilar additive process. However, the mapping $J_c^{(\kappa)}$ does not have this property, as we will see in Proposition 3.27.

Proposition 3.18 *Let $c > 0$ and $\kappa > 0$. The mappings $I_c^{(\kappa)}$ and $J_c^{(\kappa)}$ are one-to-one from $\mathbf{Ss}_c$ onto $\mathbf{Ss}_{c\kappa}$. The mapping $F^{(\kappa)}$ is one-to-one from* **St** *onto* **St**. *We have*

$$(I_c^{(\kappa)})^{-1} = I_{c\kappa}^{(1/\kappa)}, \quad (J_c^{(\kappa)})^{-1} = J_{c\kappa}^{(1/\kappa)}, \quad (F^{(\kappa)})^{-1} = F^{(1/\kappa)}, \tag{3.6}$$

for their inverses.

Proof It is enough to verify

$$I_{c\kappa}^{(1/\kappa)} \circ I_c^{(\kappa)} = J_{c\kappa}^{(1/\kappa)} \circ J_c^{(\kappa)} = \text{identity mapping of } \mathbf{Ss}_c,$$

$$I_c^{(\kappa)} \circ I_{c\kappa}^{(1/\kappa)} = J_c^{(\kappa)} \circ J_{c\kappa}^{(1/\kappa)} = \text{identity mapping of } \mathbf{Ss}_{c\kappa},$$

$$F^{(1/\kappa)} \circ F^{(\kappa)} = F^{(\kappa)} \circ F^{(1/\kappa)} = \text{identity mapping of } \mathbf{St}.$$

These are easily shown from the definitions. ■

Now we can examine Remark 3.15 more in detail.

Proposition 3.19 *Let $c > 0$ and $\kappa > 0$. Then*

$$\Lambda'_{c\kappa} \circ \Lambda_c = J_c^{(\kappa)}, \tag{3.7}$$

$$\Lambda'_{c\kappa} \circ F^{(\kappa)} \circ \Lambda_c = I_c^{(\kappa)}. \tag{3.8}$$

Proof Let $\{X_t\} \in \mathbf{Ss}_c$ and $\{Y_s\} = \Lambda_c(\{X_t\}) = \{e^{-cs} X_{\exp s}\}$. To see (3.7), note that $\Lambda'_{c\kappa}(\{Y_s\}) = \{t^{c\kappa} Y_{\log t}\} = \{t^{c\kappa} t^{-c} X_t\} = J_c^{(\kappa)}(\{X_t\})$. To see (3.8), note that $\{Y'_s\} = F^{(\kappa)}(\{Y_s\}) = \{Y_{\kappa s}\} = \{e^{-c\kappa s} X_{\exp(\kappa s)}\}$ and that $\Lambda'_{c\kappa}(\{Y'_s\}) = \{t^{c\kappa} Y'_{\log t}\} = \{t^{c\kappa} t^{-c\kappa} X_{t^\kappa}\} = I_c^{(\kappa)}(\{X_t\})$. ■

Next we show that Lamperti transformation preserves Markov property. Given the two processes $\{X_t : t \geq 0\}$ and $\{Y_s : s \in \mathbb{R}\}$ as in Theorem 3.13, let $\mathcal{F}_t^X$, $t \geq 0$, and $\mathcal{F}_s^Y$, $s \in \mathbb{R}$, be the σ-algebras generated by $\{X_{t'} : t' \leq t\}$ and by $\{Y_{s'} : s' \leq s\}$, respectively.

Theorem 3.20 *Let $c > 0$. Let $\{X_t : t \geq 0\} \in \mathbf{Ss}_c$, $\{Y_s : s \in \mathbb{R}\} \in \mathbf{St}$, and $\{Y_s\} = \Lambda_c(\{X_t\})$. Then $\mathcal{F}_s^Y = \mathcal{F}_{\exp s}^X$ for $s \in \mathbb{R}$ and $\mathcal{F}_t^X = \mathcal{F}_{\log t}^Y$ for $t > 0$. If one of the processes $\{X_t\}$ and $\{Y_s\}$ is a Markov process, then the other is also a Markov process. If $\{X_t\}$ is a Markov process with transition function $P_{t_1,t_2}^X(x, B)$, then $\{Y_s\}$ is a Markov process with transition function*

$$P_{s_1,s_2}^Y(y, B) = P_{\exp s_1,\exp s_2}^X(e^{cs_1} y, e^{cs_2} B). \tag{3.9}$$

If $\{Y_s\}$ is a Markov process with transition function $P_{s_1,s_2}^Y(y, B)$, then $\{X_t\}$ is a Markov process with transition function

$$P_{t_1,t_2}^X(x, B) = P_{\log t_1,\log t_2}^Y(t_1^{-c} x, t_2^{-c} B). \tag{3.10}$$

Proof Since $Y_s = e^{-cs} X_{\exp s}$ and $X_t = t^c Y_{\log t}$, the asserted relation of $\mathcal{F}_s^Y$ and $\mathcal{F}_t^X$ is straightforward. Suppose that $\{X_t\}$ is a Markov process with transition function $P_{t_1,t_2}^X(x, B)$ (see Definition 2.12). The defining property is, for $t_1 < t_2$,

$$P\left[\{X_{t_2} \in B\} \cap H\right] = E\left[P_{t_1,t_2}^X(X_{t_1}, B) 1_H\right] \text{ for } H \in \mathcal{F}_{t_1}^X,\ B \in \mathcal{B}(\mathbb{R}^d).$$

Let $s_1 < s_2$ and let $F(y, B) = P_{\exp s_1,\exp s_2}^X(e^{cs_1} y, e^{cs_2} B)$. Then, for $H \in \mathcal{F}_{s_1}^Y$,

$$\begin{aligned} P\left[\{Y_{s_2} \in B\} \cap H\right] &= P\left[\{X_{\exp s_2} \in e^{cs_2} B\} \cap H\right] \\ &= E\left[P_{\exp s_1,\exp s_2}^X(X_{\exp s_1}, e^{cs_2} B) 1_H\right] = E\left[F(Y_{s_1}, B) 1_H\right]. \end{aligned}$$

Hence $\{Y_s : s \in \mathbb{R}\}$ is a Markov process with transition function $P_{s_1,s_2}^Y(y, B)$ equal to $F(y, B)$. The converse assertion is proved similarly. ■

Now we are ready to consider the case where $\{X_t : t \geq 0\}$ is a selfsimilar additive process. This is related to the process defined below.

Definition 3.21 Let $\rho \in ID(\mathbb{R}^d)$ and $c > 0$. A stochastic process $\{Y_s : s \in \mathbb{R}\}$ on $\mathbb{R}^d$ is called *stationary Ornstein–Uhlenbeck type process generated by ρ and c* if it is a stationary process and, at the same time, a Markov process such that, for any $s \in \mathbb{R}$, $\{Y_{s+t} : t \geq 0\}$ is the Ornstein–Uhlenbeck type process generated by ρ and c in Sect. 2.2. Here $\mathcal{L}(Y_s)$ does not depend on s; it is called *stationary distribution* of $\{Y_s : s \in \mathbb{R}\}$.

Remark 3.22 Let $\{Y_s : s \in \mathbb{R}\}$ be a stationary OU type process generated by ρ and c. If $\{Y_s\}$ is a non-constant process (that is, if $\mathcal{L}(Y_0)$ is not a δ-distribution), then it follows from Proposition 2.16 that ρ and c are uniquely determined. In this case ρ automatically belongs to $ID_{\log}$ and $\mathcal{L}(Y_s) = \Phi_{(c)}(\rho)$ for all $s \in \mathbb{R}$. Indeed, let $\mu = \mathcal{L}(Y_s)$. For $s_1 < s_2$ and $B \in \mathcal{B}(\mathbb{R}^d)$, we have

$$\mu(B) = P[Y_{s_2} \in B] = \int_{\mathbb{R}^d} \mu(dy) P_{s_2-s_1}^Y(y, B). \tag{3.11}$$

If $\rho \in ID \setminus ID_{\log}$, then, letting $s_2 \to \infty$, we obtain from (3.11) and Theorem 2.17 (iii) that $\mu(B) = 0$ for all bounded Borel set B, which is absurd. Thus we have $\rho \in ID_{\log} = \mathfrak{D}(\Phi_{(c)})$. Letting $s_2 \to \infty$ again in (3.11) and using Theorem 2.17 (i), we see that $\mu = \Phi_{(c)}(\rho)$. Hence $\{Y_s : s \in \mathbb{R}\}$ is the unique (in law) stationary OU type process generated by $\rho \in ID_{\log}$ and $c > 0$. Conversely, for any $\rho \in ID_{\log}$ and $c > 0$ we can construct a stationary OU type process generated by ρ and c.

Theorem 3.23 *Fix $c > 0$. Let $\{X_t : t \geq 0\} \in \mathbf{Ss}_c$, $\{Y_s : s \in \mathbb{R}\} \in \mathbf{St}$, and $\{Y_s\} = \Lambda_c(\{X_t\})$. Assume that $\mathcal{L}(X_1)$ is non-trivial and that $\{Y_s\}$ is a non-constant process. Then $\{X_t : t \geq 0\}$ is a c-selfsimilar additive process in law with $\mathcal{L}(X_1) = \mu$ if and only if $\{Y_s : s \in \mathbb{R}\}$ is a stationary Ornstein–Uhlenbeck type process generated by $\rho \in ID_{\log}$ and c satisfying $\Phi_{(c)}(\rho) = \mathcal{L}(Y_s) = \mu$.*

We prepare the following fact.

Proposition 3.24

(i) Let $\{X_t : t \geq 0\}$ be an additive process in law on $\mathbb{R}^d$ with $\mathcal{L}(X_{t_1}) = \mu_{t_1}$ and $\mathcal{L}(X_{t_2} - X_{t_1}) = \mu_{t_1,t_2}$ for $0 \leq t_1 \leq t_2$. Then $\{X_t : t \geq 0\}$ is a Markov process with transition function $P^X_{t_1,t_2}(x, B) = \mu_{t_1,t_2}(B - x)$ and

$$\int_{\mathbb{R}^d} e^{i\langle z,x'\rangle} P^X_{t_1,t_2}(x, dx') = \widehat{\mu}_{t_1,t_2}(z) e^{i\langle x,z\rangle} = \frac{\widehat{\mu}_{t_2}(z)}{\widehat{\mu}_{t_1}(z)} e^{i\langle x,z\rangle}. \tag{3.12}$$

(ii) Let $\{X_t : t \geq 0\}$ be a stochastically continuous Markov process on $\mathbb{R}^d$ with initial distribution δ_0 and transition function $P^X_{t_1,t_2}(x, B)$ spatially homogeneous in the sense that $P^X_{t_1,t_2}(x, B) = P^X_{t_1,t_2}(0, B - x)$ for all x and B. Then $\{X_t : t \geq 0\}$ is an additive process in law.

Proof

(i) We have, for $0 \leq t_1 < t_2 < \cdots < t_n$,

$$X_{t_n} = X_{t_1} + (X_{t_2} - X_{t_1}) + \cdots + (X_{t_n} - X_{t_{n-1}}),$$

and the right-hand side is the sum of independent summands. It follows that $\{X_t\}$ is a Markov process with transition function $P^X_{t_1,t_2}(x, B) = \mu_{t_1,t_2}(B - x)$, since $P[X_{t_1} \in B_1, X_{t_2} \in B_2] = E[1_{B_1}(X_{t_1})\mu_{t_1,t_2}(B_2 - X_{t_1})]$. The expression (3.12) follows from this.

(ii) Define $\mu_t = \mathcal{L}(X_t)$ and μ_{t_1,t_2} for $0 \leq t_1 < t_2$ by $\mu_{t_1,t_2}(B) = P^X_{t_1,t_2}(0, B)$. Then

$$P^X_{t_1,t_2}(x, B) = P^X_{t_1,t_2}(0, B - x) = \mu_{t_1,t_2}(B - x) = \int \mu_{t_1,t_2}(dx_2) 1_B(x + x_2).$$

It follows that, for $0 \leq t_1 < t_2 < \cdots < t_n$,

$$P[X_{t_1} \in B_1, \dots, X_{t_n} \in B_n] = \int \cdots \int \mu_{t_1}(dx_1) 1_{B_1}(x_1)$$
$$\mu_{t_1,t_2}(dx_2) 1_{B_2}(x_1 + x_2) \cdots \mu_{t_{n-1},t_n}(dx_n) 1_{B_2}(x_1 + \cdots + x_n).$$

Then, we can prove that $X_{t_1}, X_{t_2} - X_{t_1}, \ldots, X_{t_n} - X_{t_{n-1}}$ are independent; see Theorem 10.4 of [93] for details. ■

Proof of Theorem 3.23 Suppose that $\{X_t : t \geq 0\}$ is a c-selfsimilar additive process in law with $\mathcal{L}(X_1) = \mu$ and μ is not a δ-distribution. Then $\mu \in L_0$ by Theorem 3.5. By Proposition 3.24, $\{X_t\}$ is a Markov process with transition function $P^X_{t_1,t_2}(x, B)$ having characteristic function $\left(\widehat{\mu}(t_2^c z)/\widehat{\mu}(t_1^c z)\right) e^{i\langle x,z\rangle}$, since $X_t \stackrel{d}{=} t^c X_1$. Hence $\{Y_s : s \in \mathbb{R}\}$ is a Markov process with transition function (3.9) in Theorem 3.20 and $\mathcal{L}(Y_s) = \mathcal{L}(Y_0) = \mu$. Let $\eta(B) = P^X_{\exp s_1, \exp s_2}(e^{cs_1} y, B)$. Then

$$\int_{\mathbb{R}^d} e^{i\langle z,y'\rangle} P^Y_{s_1,s_2}(y, dy') = \widehat{\eta}(e^{-cs_2} z) = \frac{\widehat{\mu}(e^{cs_2} e^{-cs_2} z)}{\widehat{\mu}(e^{cs_1} e^{-cs_2} z)} \exp(i\langle e^{cs_1} y, e^{-cs_2} z\rangle)$$

$$= \frac{\widehat{\mu}(z)}{\widehat{\mu}(e^{-c(s_2-s_1)} z)} \exp(i\langle e^{-c(s_2-s_1)} y, z\rangle)$$

$$= \exp\left[i\langle e^{-c(s_2-s_1)} y, z\rangle + \int_0^\infty \psi_\rho(e^{-cu} z) du - \int_0^\infty \psi_\rho(e^{-cu-c(s_2-s_1)} z) du \right]$$

$$= \exp\left[i\langle e^{-c(s_2-s_1)} y, z\rangle + \int_0^{s_2-s_1} \psi_\rho(e^{-cu} z) du \right].$$

In the above we have used Theorem 2.17 (ii) for $\mu \in L_0$ and chosen $\rho \in ID_{\log}$ such that $\widehat{\mu}(z) = \exp \int_0^\infty \psi_\rho(e^{-cu} z) du$. The formula that we have obtained coincides with (2.20). Hence $\{Y_s\}$ is a non-constant stationary OU type process generated by ρ and c with stationary distribution μ.

Conversely, suppose that $\{Y_s : s \in \mathbb{R}\}$ is a non-constant stationary OU type process generated by $\rho \in ID_{\log}$ and c with $\Phi_{(c)}(\rho) = \mathcal{L}(Y_s) = \mu$. We know that $\{X_t\}$ is c-selfsimilar and $\mathcal{L}(X_1) = \mathcal{L}(Y_0) = \mu$. All we have to show is that $\{X_t\}$ is an additive process. Since $\{Y_s\} = \Lambda_c(\{X_t\})$, we know from Theorem 3.20 that $\{X_t\}$ is a Markov process and (3.9) and (3.10) hold. We claim that $P^X_{t_1,t_2}(x, B) = P^X_{t_1,t_2}(0, B - x)$ for all t_1, t_2, x, and B. Let $\zeta(B) = P^Y_{\log t_1, \log t_2}(t_1^{-c} x, B)$. Then it follows from (3.10) that $\int e^{i\langle z,x'\rangle} P^X_{t_1,t_2}(x, dx') = \widehat{\zeta}(t_2^c z)$. We know, from (2.20),

$$\widehat{\zeta}(z) = \exp\left[i\langle e^{-c(\log t_2 - \log t_1)} t_1^{-c} x, z\rangle + \int_0^{\log t_2 - \log t_1} \psi_\rho(e^{-cu} z) du \right].$$

Hence

$$\int e^{i\langle z,x'\rangle} P^X_{t_1,t_2}(x, dx') = \exp\left[i\langle x, z\rangle + \int_0^{\log t_2 - \log t_1} \psi_\rho(e^{-c(u-\log t_2)} z) du \right].$$

It follows that

$$\int e^{i\langle z,x'\rangle} P^X_{t_1,t_2}(x, dx') = e^{i\langle x,z\rangle} \int e^{i\langle z,x'\rangle} P^X_{t_1,t_2}(0, dx').$$

This means that $P^X_{t_1,t_2}(x, B) = P^X_{t_1,t_2}(0, B - x)$ as we claimed. Now, using Proposition 3.24 (ii), we conclude that $\{X_t\}$ is an additive process in law. ■

Corollary 3.25 *Any stationary OU type process $\{Y_s : s \in \mathbb{R}\}$ on $\mathbb{R}^d$ has stationary distribution $\mathcal{L}(Y_s)$ in L_0. Conversely, any distribution in $L_0(\mathbb{R}^d)$ is the stationary distribution of some stationary OU type process.*

Let us prove the following additional results.

Proposition 3.26 *Let $\{Y_s\}$ be a non-constant stationary OU type process generated by some $\rho \in ID_{\log}$ and $c > 0$ and let $\{X'_t\} = \Lambda'_{c'}(\{Y_s\})$ with $c' > 0$ different from c. Then $\{X'_t\}$ is a c'-selfsimilar process with non-trivial $\mathcal{L}(X_1)$, but $\{X_t\}$ is not an additive process in law.*

Proof By Theorem 3.13, $\{X'_t\}$ is a c'-selfsimilar process with non-trivial $\mathcal{L}(X'_1)$ and $\{Y_s\} = \Lambda_{c'}(\{X'_t\})$. Suppose that $\{X'_t\}$ is an additive process in law. Then, by Theorem 3.23, $\{Y_s\}$ is a non-constant stationary OU type process generated by $\Phi^{-1}_{(c')}(\mu)$ and c', where $\mu = \mathcal{L}(Y_0)$. On the other hand, $\{Y_s\}$ is a stationary OU type process generated by $\rho = \Phi^{-1}_{(c)}(\mu)$ and c. This is contradictory to the uniqueness of generator in Proposition 2.16 (i). Therefore $\{X'_t\}$ is not an additive process in law. ■

Proposition 3.27 *Let $\{X_t\}$ be a c-selfsimilar additive process in law with non-trivial $\mathcal{L}(X_1)$. Let $\{X'_t\} = J^{(\kappa)}_c(\{X_t\})$ with $\kappa \in (0, 1) \cup (1, \infty)$, where $J^{(\kappa)}_c$ is the mapping in Definition 3.17. Then $\{X'_t\}$ is a $c\kappa$-selfsimilar process with $\mathcal{L}(X'_1) = \mathcal{L}(X_1)$, but $\{X'_t\}$ is not an additive process in law.*

Proof This is another expression of Proposition 3.26. Apply that proposition to $\{Y_s\} = \Lambda_c(\{X_t\})$ and $\{X'_t\} = \Lambda'_{c\kappa}(\{Y_s\})$ and notice $\Lambda'_{c\kappa} \circ \Lambda_c = J^{(\kappa)}_c$ in Proposition 3.19. ■

Example 3.28 Let us directly check Proposition 3.27 for Brownian motion $\{X_t\}$ on $\mathbb{R}$. Recalling that $\{X_t\}$ is (1/2)-selfsimilar, let $\{X'_t\} = J^{(\kappa)}_{1/2}(\{X_t\}) = \{t^{(\kappa-1)/2}X_t\}$ with $\kappa \in (0, 1) \cup (1, \infty)$. In order to prove that $\{X'_t\}$ is not an additive process in law, it is enough to show that X'_1 and $X'_2 - X'_1$ are not independent. Let us check variances. We have $\mathrm{var}X'_1 = \mathrm{var}X_1 = 1$ and

$$\begin{aligned}\mathrm{var}(X'_2 - X'_1) &= \mathrm{var}(2^{(\kappa-1)/2}X_2 - X_1)\\ &= \mathrm{var}(2^{(\kappa-1)/2}(X_2 - X_1) + (2^{(\kappa-1)/2} - 1)X_1)\\ &= 2^{\kappa-1} + (2^{(\kappa-1)/2} - 1)^2 = 2^{\kappa} - 2^{(\kappa-1)/2+1} + 1.\end{aligned}$$

If X'_1 and $X'_2 - X'_1$ are independent, then

$$\mathrm{var}X'_2 = \mathrm{var}X'_1 + \mathrm{var}(X'_2 - X'_1) = 2^{\kappa} - 2^{(\kappa-1)/2+1} + 2,$$

which is contradictory to $\mathrm{var}X'_2 = \mathrm{var}(2^{(\kappa-1)/2}X_2) = 2^{\kappa}$.

Remark 3.29 Let $\{X_t : t \geq 0\}$ and $\{Y_s : s \in \mathbb{R}\}$ be a non-trivial c-selfsimilar additive process in law and a non-constant stationary OU type process generated by $\rho \in ID_{\log}$ and c, respectively, with $\{Y_s\} = \Lambda_c(\{X_t\})$. Then the background driving Lévy process $\{Z_u\} = \{Z_u^{(\rho)}\}$ of $\{Y_s\}$ is constructed from $\{X_t\}$ in the following way. Fix an arbitrary $r \in \mathbb{R}$. Define $Z_u = \int_{\exp r}^{\exp(r+u)} t^{-c} dX_t$ for $u \geq 0$, where the stochastic integral based on an additive process on $\mathbb{R}^d$ is defined similarly to Definition 2.3 (see Sato [96] (2004)). Then $\{Z_u : u \geq 0\}$ is a Lévy process. Indeed, it is clearly an additive process in law and, moreover, for $0 \leq u_1 < u_2$ and $p > 0$, $Z_{u_2+p} - Z_{u_1+p} = \int_{\exp(r+u_1+p)}^{\exp(r+u_2+p)} t^{-c} dX_t = \int_{\exp(r+u_1)}^{\exp(r+u_2)} (at)^{-c} dX_{at} \stackrel{d}{=} \int_{\exp(r+u_1)}^{\exp(r+u_2)} t^{-c} dX_t = Z_{u_2} - Z_{u_1}$ with $a = e^p$, since $\{a^{-c} X_{at}\} \stackrel{d}{=} \{X_t\}$. We can prove, for $0 \leq u_1 \leq u_2$, $\int_{u_1}^{u_2} f(u) dZ_u = \int_{\exp(r+u_1)}^{\exp(r+u_2)} f(\log t - r) t^{-c} dX_t$ for all locally bounded measurable functions f on $[0, \infty)$, since this holds for step functions. Hence $\int_0^s e^{cu} dZ_u = \int_{\exp r}^{\exp(r+s)} e^{-cr} dX_u = e^{-cr}(X_{\exp(r+s)} - X_{\exp r})$. It follows that $Y_{r+s} = e^{-cs} Y_r + e^{-cs} \int_0^s e^{cu} dZ_u$ for $s \geq 0$, which is a special case of (2.14), showing that $\{Z_u\}$ is the background driving Lévy process of the OU type process $\{Y_{r+s} : s \geq 0\}$. Note that Y_r and $\{Z_u\}$ are independent.

Notes

Selfsimilar additive processes were studied by Sato [88, 89] (1990, 1991) for the first time. Theorem 3.5 was proved in [89] in a more general form called wide-sense operator selfsimilar additive processes, which are related to Q-selfdecomposable distributions in Sect. 5.2. Thereafter selfsimilar additive processes are employed as a model in mathematical finance under the name Sato processes, such as in Carr et al. [17] (2005), Eberlein and Madan [23] (2009), and Kokholm and Nicolato [48] (2010). Theorem 3.8 was given by Maejima et al. [62] (2000).

Path behaviours of selfsimilar additive processes are studied by Watanabe [132] (1996), Sato and Yamamuro [106, 107] (1998, 2000), Yamamuro [141, 142] (2000a, 2000b), and Watanabe and Yamamuro [137] (2010). One of the results is that any nondegenerate selfsimilar additive process on $\mathbb{R}^d$ is transient if $d \geq 3$, like any nondegenerate Lévy process.

Lemma 3.9 (i) says that any linear transformation of a distribution in L_m belongs to L_m. In addition to this fact, there is $\mu = \mathcal{L}(X) \in L_m(\mathbb{R}^d) \setminus L_{m+1}(\mathbb{R}^d)$ such that, for every $d' \times d$ real matrix F with $1 \leq d' < d$, $\mathcal{L}(FX)$ is in $L_{m+1}(\mathbb{R}^{d'})$. This statement is true for $m = -2, -1, 0, 1, \ldots < \infty$ with understanding that $L_{-1} = ID$ and $L_{-2} = \mathfrak{P}$. This is by Sato [92] (1998) for $m = -1$ and by Maejima et al. [63] (1999) for $m \geq 0$. For $m = -2$, see the references in [92]; an explicit construction of μ using "signed Lévy measure" is done.

The two representations of selfdecomposable distributions in selfsimilar additive processes (Theorem 3.5) and in stationary OU type processes (Corollary 3.25) were originally proved separately. The former edition of this book constructed stationary OU type processes with time parameter in $(-\infty, \infty)$ by stochastic integrals. But those two kinds of processes are connected by a transformation introduced by Lamperti [53] (1962), which is now called Lamperti transformation. This fact was pointed out by Jeanblanc, Pitman, and Yor [38] (2002) and examples induced by Bessel processes were examined. Maejima and Sato [60] (2003) extended this connection to the processes called semi-selfsimilar additive and semi-stationary OU type in the representations of the semi-selfdecomposable distributions. In the present edition we have made a close inspection of Lamperti transformation and constructed stationary OU type processes via this transformation.

As a typical non-stable example, let μ be the exponential distribution on $\mathbb{R}$ with mean 1 and let $\{Y_t\}$ and $\{X_t\}$ be the Lévy process and the 1-selfsimilar additive process, respectively, with $\mathcal{L}(Y_1) = \mathcal{L}(X_1) = \mu$. Then, the jump times of $Y_t(\omega)$ are countable and dense in $(0, \infty)$ a. s., while the path of $X_t(\omega)$ is a step function of $t \in (\varepsilon, \infty)$ for any $\varepsilon > 0$ a. s. and the jump times of $X_t(\omega)$ accumulate only at $t = 0$ a. s. Moreover, $\lim_{t\downarrow 0} t^{-a} Y_t(\omega) = 0$ a. s. for any $a > 0$, while $\limsup_{t\downarrow 0} X_t(\omega)/(t \log|\log t|) = 1$ a. s. Thus path behaviours of $\{X_t\}$ and $\{Y_t\}$ are greatly different. See Sato [89] (1991).

An active area related to selfsimilarity is the study of *positive selfsimilar Markov processes* and their connection with Lévy processes. See Chapter 13 of Kyprianou [51] (2014) for definition and examples.

Chapter 4
Multivariate Subordination

Subordination of stochastic processes consists of transforming a stochastic process $\{X_t\}$ to another one through random time change by an increasing process $\{Z_t\}$, where $\{X_t\}$ and $\{Z_t\}$ are assumed to be independent.

Subordination of a Lévy process $\{X_t\}$ on $\mathbb{R}^d$ by an increasing Lévy process $\{Z_t\}$ on $\mathbb{R}$ is studied in books such as Feller [24] (1971) and Sato [93] (1999). In this case subordination provides a Lévy process; it means introducing a new Lévy process $\{Y_t\}$ by defining $Y_t = X_{Z_t}$. In Sect. 4.1 we recall the Lévy–Khintchine representation of the characteristic function of an increasing Lévy process and the description of the generating triplet of $\{Y_t\}$ in terms of those of $\{X_t\}$ and $\{Z_t\}$. Further, characterization of Lévy processes on a cone K in $\mathbb{R}^N$ is given. They are called K-valued subordinators, the extended notion of an increasing Lévy process.

In Sect. 4.2, the concept of a Lévy process is expanded by replacing the time parameter t in $[0, \infty)$ by a parameter s in a cone K in $\mathbb{R}^N$. A natural generalization of the stationary independent increment property by assuming stationary independent increments along K-increasing sequences leads to the notion of a K-parameter Lévy process on $\mathbb{R}^d$. Certain continuity conditions are added in the definition. The concept of subordination is extended to substitution of s by a K-valued subordinator $\{Z_t\}$. It is proved that $Y_t = X_{Z_t}$ is a Lévy process.

The positive orthant $\mathbb{R}_+^N$ is a cone. A deeper study in the case $K = \mathbb{R}_+^N$ is made in Sect. 4.3. Joint distributions of $\{X_s : s \in K\}$ are examined and the relations of the generating triplets involved in subordination are clarified. The case of general cones K is examined in Sect. 4.4.

The notion of cone-parameter convolution semigroups $\{\mu_s : s \in K\}$ on $\mathbb{R}^d$ is introduced in Sect. 4.2. Here a K-parameter convolution semigroup is a family of distributions on $\mathbb{R}^d$ that satisfies $\mu_{s^1+s^2} = \mu_{s^1} * \mu_{s^2}$ for $s^1, s^2 \in K$ and has a

A. Rocha-Arteaga, K. Sato, *Topics in Infinitely Divisible Distributions and Lévy Processes*, SpringerBriefs in Probability and Mathematical Statistics,
https://doi.org/10.1007/978-3-030-22700-5_4

continuity property. In Sect. 4.4 the problem is discussed whether or not a given K-parameter convolution semigroup generates a K-parameter Lévy process and whether uniquely or not if it generates. The situation is complicated but we touch on typical results and examples. Further, subordination of cone-parameter convolution semigroups is presented.

4.1 Subordinators and Subordination

Basic results on subordination of a Lévy process on $\mathbb{R}^d$ by an increasing Lévy process on $\mathbb{R}$ are presented here but their proofs are omitted (they can be found in the book [93]). Increasing Lévy processes on $\mathbb{R}$ are often called subordinators. Further, in this section, we consider K-valued subordinators, that is, K-valued (or K-increasing) Lévy processes.

Definition 4.1 A Lévy process $\{Z_t : t \geq 0\}$ on $\mathbb{R}$ is called *increasing* if $Z_t(\omega)$ is increasing in t, almost surely. An increasing Lévy process on $\mathbb{R}$ is called a *subordinator*.

Theorem 4.2 *A Lévy process* $\{Z_t : t \geq 0\}$ *on* $\mathbb{R}$ *with generating triplet* (A_Z, ν_Z, γ_Z) *is a subordinator if and only if*

$$A_Z = 0, \ \nu_Z((-\infty, 0)) = 0, \int_{(0,1]} x\nu_Z(dx) < \infty, \ \textit{and} \ \gamma_Z^0 \geq 0, \tag{4.1}$$

where $\gamma_Z^0 = \gamma_Z - \int_{(0,1]} x\nu_Z(dx)$ *is the drift of* $\{Z_t\}$*. For any* $w \in \mathbb{C}$ *with* $\mathrm{Re}\, w \leq 0$,

$$E[e^{wZ_t}] = e^{t\Psi(w)}, \tag{4.2}$$

$$\Psi(w) = \gamma_Z^0 w + \int_{(0,\infty)} (e^{ws} - 1)\nu_Z(ds). \tag{4.3}$$

For a proof see [93, Theorem 21.5]. Notice that (4.2)–(4.3) represent characteristic function if $w = iz$, $z \in \mathbb{R}$, and Laplace transform if $w = -u$, $u \geq 0$.

Theorem 4.3 *Let* $\{Z_t : t \geq 0\}$ *be a subordinator with Lévy measure* ν_Z*, drift* γ_Z^0*, and* $\Psi(w)$ *of (4.3). Let* $\{X_t : t \geq 0\}$ *be a Lévy process on* $\mathbb{R}^d$ *with generating triplet* (A_X, ν_X, γ_X) *and* $\mu = \mathcal{L}(X_1)$*. Assume that* $\{Z_t\}$ *and* $\{X_t\}$ *are independent. Define* $Y_t = X_{Z_t}$*. Then* $\{Y_t : t \geq 0\}$ *is a Lévy process on* $\mathbb{R}^d$ *and*

$$Ee^{i\langle z, Y_t\rangle} = e^{t\Psi((\log\widehat{\mu})(z))}. \tag{4.4}$$

The generating triplet (A_Y, ν_Y, γ_Y) *of* $\{Y_t\}$ *is as follows:*

$$A_Y = \gamma_Z^0 A_X, \tag{4.5}$$

$$\nu_Y(B) = \int_{(0,\infty)} \mu^s(B)\nu_Z(ds) + \gamma_Z^0 \nu_X(B) \quad \textit{for } B \in \mathcal{B}(\mathbb{R}^d \setminus \{0\}), \tag{4.6}$$

$$\gamma_Y = \int_{(0,\infty)} \nu_Z(ds) \int_{|x|\le 1} x\mu^s(dx) + \gamma_Z^0 \gamma_X. \tag{4.7}$$

If $\gamma_Z^0 = 0$ *and* $\int_{(0,1]} s^{1/2}\nu_Z(ds) < \infty$, *then* $A_Y = 0$, $\int_{|x|\le 1} |x|\nu_Y(dx) < \infty$, *and the drift of* $\{Y_t\}$ *is zero.*

This is Theorem 30.1 of [93]. It is a special case of Theorems 4.23 and 4.41, of which we will give a full proof.

Definition 4.4 The procedure in Theorem 4.3 of making $\{Y_t\}$ from $\{Z_t\}$ and $\{X_t\}$ is called (Bochner's) *subordination*. We say that $\{Y_t\}$ is *subordinate* to $\{X_t\}$ by $\{Z_t\}$. Sometimes we call $\{X_t\}$ *subordinand* and $\{Y_t\}$ *subordinated.*

In the proof of Theorem 4.3 the following fact is essential.

Lemma 4.5 *Let* $\{X_t\}$ *be a Lévy process on* $\mathbb{R}^d$. *Then there are constants* $C(\varepsilon)$, C_1, C_2, C_3 *such that*

$$P[|X_t| > \varepsilon] \le C(\varepsilon)t \quad \textit{for } \varepsilon > 0,$$

$$E[|X_t|^2; |X_t| \le 1] \le C_1 t,$$

$$|E[X_t; |X_t| \le 1]| \le C_2 t,$$

$$E[|X_t|; |X_t| \le 1] \le C_3 t^{1/2}.$$

This is Lemma 30.3 of [93]. It is a special case of $d = N = 1$ in Lemma 4.40.

A stochastic process $\{X_t\}$ on $\mathbb{R}^d$ is called *rotation invariant* if $\{X_t\} \stackrel{d}{=} \{UX_t\}$ for every $d \times d$ orthogonal matrix U.

Example 4.6 Let $\{X_t\}$ be the Brownian motion on $\mathbb{R}^d$ and $\{Z_t\}$ a non-trivial α-stable subordinator, $0 < \alpha < 1$, in Theorem 4.3.

By Theorems 4.2 and 1.42, $\{Z_t\}$ has characteristic function

$$Ee^{i\langle z, Z_1\rangle} = \exp\left[c_1 \int_{(0,\infty)} (e^{izx} - 1)x^{-1-\alpha}dx + i\gamma^0 z\right]$$

with $0 < \alpha < 1$, $\gamma^0 \ge 0$, and $c_1 > 0$. Or, equivalently, we can write

$$Ee^{i\langle z, Z_1\rangle} = \exp\left[-c\,|z|^\alpha \left(1 - i\tan\left(\frac{\pi\alpha}{2}\,\mathrm{sgn}\,(z)\right)\right) + i\gamma^0 z\right]$$

with $c = c_1\alpha^{-1}\Gamma(1-\alpha)\cos(\pi\alpha/2) > 0$. Thus an α-stable process on $\mathbb{R}$ with parameter $(\alpha, \kappa, \tau, c)$ of Definition 14.16 of [93] is a subordinator if and only if $0 < \alpha < 1$, $\beta = 1$, $\tau = \gamma^0 \geq 0$. The function $\Psi(w)$ in (4.3) for $w = -u \leq 0$ is given by

$$\Psi(-u) = -c'u^\alpha - \gamma^0 u$$

with $c' = c_1\alpha^{-1}\Gamma(1-\alpha)$ [93, Example 24.12]. Thus, $\{Z_t\}$ is a non-trivial strictly α-stable subordinator if and only if, in addition, $\gamma^0 = 0$; see Theorem 1.42 (iii).

Now let $\{Z_t\}$ be a non-trivial strictly α-stable subordinator. Then $\Psi(-u) = -c'u^\alpha$. From $\log\widehat{\mu}(z) = -(1/2)|z|^2$, we get

$$Ee^{i\langle z, Y_t\rangle} = \exp\left[-\frac{t}{2^\alpha}c'|z|^{2\alpha}\right]$$

by (4.4). Hence, $\{Y_t\}$ is a rotation invariant 2α-stable process. See [93], Theorem 14.14, for a characterization of a rotation invariant stable distribution.

Example 4.7 Let, in Theorem 4.3, $\{X_t\}$ be a Lévy process on $\mathbb{R}^d$ and $\{Z_t\}$ Γ-process with parameter $q > 0$. Define a measure V^q on $\mathbb{R}^d$ by $\mathcal{L}(Y_1) = (1/q)V^q$. Then V^q is called the *q-potential measure* of $\{X_t\}$.

We have

$$\Psi(-u) = \int_0^\infty \left(e^{-ux} - 1\right)\frac{e^{-qx}}{x}dx = -\log\left(1 + \frac{u}{q}\right), \quad u \geq 0,$$

see Example 1.45. The subordination $Y_t = X_{Z_t}$ gives

$$Ee^{i\langle z, Y_t\rangle} = e^{t[-\log(1-q^{-1}\log\widehat{\mu}(z))]} = \left(1 - q^{-1}\log\widehat{\mu}(z)\right)^{-t}, \quad z \in \mathbb{R}^d,$$

$$P(Y_t \in B) = \frac{q^t}{\Gamma(t)}\int_0^\infty P(X_s \in B)\, s^{t-1}e^{-qs}ds.$$

Hence $\mathcal{L}(Y_1) = q^{-1}V^q$ where $V^q = \int_0^\infty \mu^s(B)e^{-qs}ds$.

In particular, if $d = 1$ and $\{X_t\}$ is a Poisson process with parameter $c > 0$, then, for each $t > 0$, $Ee^{-uX_t} = e^{tc(e^{-u}-1)}$ and

$$Ee^{-uY_t} = e^{-t\log(1-q^{-1}c(e^{-u}-1))} = p^t\left(1 - (1-p)e^{-u}\right)^{-t}, \quad u \geq 0,$$

where $p = q/(c+q)$. Hence Y_t has negative binomial distribution with parameters t and $q/(c+q)$.

If $d = 1$ and $\{X_t\}$ is a symmetric α-stable process with $Ee^{izX_t} = e^{-t|z|^\alpha}$, $0 < \alpha \leq 2$, then

$$Ee^{izY_1} = \left(1 + q^{-1} |z|^{\alpha}\right)^{-1}, \quad z \in \mathbb{R}.$$

Here $\mathcal{L}(Y_1)$ is called *Linnik distribution* or *geometric stable distribution*.

Definition 4.8 A subset K of $\mathbb{R}^N$ is a *cone* if it is a nonempty closed convex set, is not $\{0\}$, and satisfies

(a) if $s \in K$ and $a \geq 0$, then $as \in K$,
(b) if $s \in K$ and $-s \in K$, then $s = 0$.[1]

Notice that a cone is closed under multiplication by nonnegative reals and does not contain any straight line through 0. Assume, in the following of this section, that K is a cone in $\mathbb{R}^N$. Then it determines a partial order.

Definition 4.9 Write $s^1 \leq_K s^2$ if $s^2 - s^1 \in K$. A sequence $\{s^n\}_{n=1,2,\ldots} \subset \mathbb{R}^N$ is *K-increasing* if $s^n \leq_K s^{n+1}$ for each n; *K-decreasing* if $s^{n+1} \leq_K s^n$ for each n. A mapping f from $[0, \infty)$ into $\mathbb{R}^N$ is *K-increasing* if $f(t_1) \leq_K f(t_2)$ for $t_1 < t_2$; *K-decreasing* if $f(t_2) \leq_K f(t_1)$ for $t_1 < t_2$.

Lemma 4.10 *A cone K has the following properties.*

(i) If $s^1 \in K$ and $s^2 \in K$, then $s^1 + s^2 \in K$.
(ii) K does not contain any straight line.
(iii) There is an $(N-1)$-dimensional linear subspace H of $\mathbb{R}^N$ such that, for any $s \in K$, $(s + H) \cap K$ is a bounded set.
(iv) If $\{s^n\}_{n=1,2,\ldots}$ is a K-decreasing sequence in K, then it is convergent.

Proof

(i) Notice that $s^1 + s^2 = 2(\frac{1}{2}s^1 + \frac{1}{2}s^2)$.
(ii) Suppose that a straight line $\{s^0 + as^1 : a \in \mathbb{R}\}$, $s^1 \neq 0$, is contained in K. Then $K \ni \frac{1}{n}(s^0 + ns^1) \to s^1$. Hence $s^1 \in K$. Similarly $-s^1 \in K$, since $K \ni \frac{1}{n}(s^0 - ns^1) \to -s^1$. Hence K contains the straight line $\{as^1 : a \in \mathbb{R}\}$, contradicting that K is a cone.
(iii) Let us admit the fact that there is an $(N-1)$-dimensional linear subspace H of $\mathbb{R}^N$ such that $K \cap H = \{0\}$ (this fact is evident if $N = 1$ or 2 or if $K = \mathbb{R}^N_+$; in general case, books in convex analysis (e. g. Rockafellar [82] (1970)) are helpful in giving a proof). We can choose $\gamma \neq 0$ such that $H = \{u : \langle u, \gamma \rangle = 0\}$ and $K \setminus \{0\} \subset \{u : \langle u, \gamma \rangle > 0\}$. We claim that $(s + H) \cap K$ is bounded for any $s \in K$. Suppose that there are $s^n \in (s + H) \cap K$ with $|s^n| \to \infty$. Then, $\langle s^n, \gamma \rangle > 0$ and $\langle s^n - s, \gamma \rangle = 0$. A subsequence of $|s^n|^{-1}s^n$ tends to some point $u \in K$ with $|u| = 1$. From $\langle |s^n|^{-1}s^n - |s^n|^{-1}s, \gamma \rangle = 0$ we have $\langle u, \gamma \rangle = 0$, which contradicts that $K \cap H = \{0\}$.

[1] In the previous edition this is called "proper cone" as the condition (b) is imposed. Here we follow [71] and [72].

(iv) We use H and γ in the proof of (iii). Let $\{s^n\}_{n=1,2,...}$ be a K-decreasing sequence in K. Let $K_1 = \{u: u \in K \text{ and } \langle u - s^1, \gamma\rangle \le 0\}$. Then K_1 is bounded. Indeed, if there are $u^n \in K_1$ with $|u^n| \to \infty$, then a limit point v of $|u^n|^{-1}u^n$ satisfies $|v| = 1$, $v \in K$, and $\langle v, \gamma\rangle \le 0$, which is absurd. Now let us show that $\{s^n\}$ is bounded. If $|s^n| \to \infty$, then $|s^1 + s^n| \to \infty$, $|s^1 - s^n| \to \infty$, and $s^1 + s^n, s^1 - s^n \in K$, and hence, for all large n, $s^1 + s^n \notin K_1$ and $s^1 - s^n \notin K_1$, which means that $\langle s^n, \gamma\rangle > 0$ and $\langle -s^n, \gamma\rangle > 0$, a contradiction. Similarly, if a subsequence $\{s^{n(k)}\}$ of $\{s^n\}$ satisfies $|s^{n(k)}| \to \infty$, we have a contradiction. Hence $\{s^n\}$ is bounded. If two subsequences $\{s^{n(k)}\}$ and $\{s^{m(l)}\}$ tend to u and v, respectively, then $v - u \in K$ since $s^{m(l)} - s^{n(k)} \in K$ for $n(k) > m(l)$, and similarly $u - v \in K$, which shows $u = v$ since K does not contain any straight line through 0. Therefore $\{s^n\}$ is convergent. ■

Let $(\Theta, \mathcal{B}, \rho)$ be a σ-finite measure space. We say that a collection $\{N(B): B \in \mathcal{B}\}$ of random variables on $\mathbb{Z}_+ \cup \{+\infty\}$ is *Poisson random measure* on Θ with *intensity measure* ρ if the following are satisfied: (a) for all B, $N(B)$ has Poisson distribution with mean $\rho(B)$, (b) if $B_1, \ldots, B_n$ are disjoint then $N(B_1), \ldots, N(B_n)$ are independent, and (c) for every ω, $N(\cdot, \omega)$ is a measure on Θ. Here we understand that a random variable X has Poisson distribution with mean 0 if $X = 0$ a. s.; X has Poisson distribution with mean ∞ if $X = \infty$ a. s.

Now let us extend Theorem 4.2 to higher dimensions.

Theorem 4.11 *Let $\{Z_t: t \ge 0\}$ be a Lévy process on $\mathbb{R}^N$ with generating triplet (A, ν, γ). Then the following three conditions are equivalent.*

(a) For any fixed $t \ge 0$, $Z_t \in K$ a. s.
(b) Almost surely, $Z_t(\omega)$ is K-increasing in t.
(c) The generating triplet satisfies

$$A = 0, \quad \nu(\mathbb{R}^N \setminus K) = 0, \quad \int_{|x|\le 1} |x|\, \nu(dx) < \infty, \quad \text{and } \gamma^0 \in K, \tag{4.8}$$

where γ^0 is the drift of $\{Z_t\}$.

Proof First, let us check the equivalence of (a) and (b). If (b) holds, then $Z_t = Z_t - Z_0 \in K$ a. s. and (a) holds. If (a) holds, then, for $0 \le s \le t$, $P[Z_t - Z_s \in K] = P[Z_{t-s} \in K] = 1$, hence

$$P[Z_t - Z_s \in K \text{ for all } s, t \text{ in } \mathbb{Q} \cap [0, \infty) \text{ with } s \le t] = 1,$$

and thus (b) holds by right continuity of sample functions and by closedness of K.

Let us show that (c) implies (a). Assume (c). By the Lévy–Itô decomposition of sample functions in Theorem 19.3 of [93],

$$Z_t(\omega) = \lim_{n\to\infty} \int_{(0,t]\times\{|x|>1/n\}} x J(d(s,x),\omega) + t\gamma^0 \quad \text{a. s.},$$

where, for $B \in \mathcal{B}\big((0,\infty)\times(\mathbb{R}^N\setminus\{0\})\big)$, $J(B,\omega)$ is defined to be the number of s such that $(s, Z_s(\omega) - Z_{s-}(\omega)) \in B$. It is known that $J(B)$ is a Poisson random measure with intensity measure being the product of Lebesgue measure on $(0,\infty)$ and ν on $\mathbb{R}^N\setminus\{0\}$. Here we have used $A = 0$ and $\int_{|x|\le 1} |x|\nu(dx) < \infty$. It follows from $\nu(\mathbb{R}^N\setminus K) = 0$ that

$$E\left[\int_{(0,t]\times(\{|x|>1/n\}\setminus K)} J(d(s,x))\right] = t\nu(\{|x|>1/n\}\setminus K) = 0$$

and hence $\int_{(0,t]\times\{|x|>1/n\}} xJ(d(s,x),\omega)$ is the sum of a finite number of x in K for each ω. This, combined with $\gamma^0 \in K$ and with (i) of Lemma 4.10 and closedness of K, shows that $Z_t \in K$ a. s.

Conversely, assume (b) and let us show (c). We use a part of the general Lévy–Itô decomposition. Since all jumps $Z_s - Z_{s-}$ are in K, we have

$$\nu(\mathbb{R}^N\setminus K) = E\left[J\Big((0,1]\times(\mathbb{R}^N\setminus K)\Big)\right] = 0.$$

We deal with ω such that $Z_t(\omega)$ is K-increasing in t and $Z_0(\omega) = 0$, and we omit ω. If $0 \le s < t$, then $Z_{t-} - Z_s = \lim_{\varepsilon\downarrow 0} Z_{t-\varepsilon} - Z_s \in K$. Hence, if $0 < s_1 < \cdots < s_n \le t$, then

$$Z_t - \sum_{k=1}^{n}(Z_{s_k} - Z_{s_k-}) = Z_t - Z_{s_n} + \sum_{k=2}^{n}(Z_{s_k-} - Z_{s_{k-1}}) + Z_{s_1} \in K.$$

Let $Z_t^{(n)} = \int_{(0,t]\times\{|x|>1/n\}} xJ(d(s,x))$. It is the sum of jumps with size bigger than $1/n$ up to time t. It follows that $Z_t - Z_t^{(n)} \in K$ and that

$$(Z_t - Z_t^{(n)}) - (Z_t - Z_t^{(n+1)}) = Z_t^{(n+1)} - Z_t^{(n)} \in K,$$

that is, $Z_t - Z_t^{(n)}$ is a K-decreasing sequence in K. Hence, by (iv) of Lemma 4.10, $Z_t - Z_t^{(n)}$ is convergent. Define $Z_t^1 = \lim_{n\to\infty} Z_t^{(n)}$ and $Z_t^2 = Z_t - Z_t^1$. We see that Z_t^1 and Z_t^2 take values in K. We claim that

$$x^n = \int_{1/n<|x|\le 1} x\nu(dx) \text{ is convergent as } n\to\infty. \tag{4.9}$$

Using Proposition 19.5 of [93], we have

$$Ee^{i\langle z, Z_t^{(n)}\rangle} = \exp\left[t\int_{|x|>1/n}(e^{i\langle z,x\rangle}-1)\nu(dx)\right]$$

$$= \exp\left[t\left(\int_{1/n<|x|\le 1}(e^{i\langle z,x\rangle}-1-i\langle z,x\rangle)\nu(dx)\right.\right.$$

$$\left.\left.+\int_{|x|>1}(e^{i\langle z,x\rangle}-1)\nu(dx)+i\int_{1/n<|x|\le 1}\langle z,x\rangle\nu(dx)\right)\right].$$

As $n\to\infty$, $Ee^{i\langle z, Z_t^{(n)}\rangle}\to Ee^{i\langle z, Z_t^1\rangle}$ and

$$\exp\left[t\int_{1/n<|x|\le 1}(e^{i\langle z,x\rangle}-1-i\langle z,x\rangle)\nu(dx)\right]$$

$$\to \exp\left[t\int_{|x|\le 1}(e^{i\langle z,x\rangle}-1-i\langle z,x\rangle)\nu(dx)\right],$$

both uniformly in z in any compact set. Hence $\exp\left[it\int_{1/n<|x|\le 1}\langle z,x\rangle\nu(dx)\right]$ is convergent uniformly in z in any compact set. That is, δ_{x^n} is convergent and, equivalently, (4.9). The meaning of (4.9) is that, componentwise, $\int_{1/n<|x|\le 1}x_j\nu(dx)$ is convergent for $j = 1,\dots,N$. Starting from $Z_t^{(n),j} = \int_{(0,t]\times\{|x|>1/n,\, x_j\ge 0\}}xJ(d(s,x))$, we can see, in the same way, that the integral $\int_{\{1/n<|x|\le 1,\, x_j\ge 0\}}x_j\nu(dx)$ is convergent as $n\to\infty$. Hence the integral $\int_{\{1/n<|x|\le 1,\, x_j<0\}}x_j\nu(dx)$ is also convergent. It follows that $\int_{1/n<|x|\le 1}|x_j|\nu(dx)$ is convergent. Hence $\int_{|x|\le 1}|x|\nu(dx)<\infty$. We can now apply Theorem 19.3 of [93] and obtain that Z_t^2 is a Lévy process with triplet $(A, 0, \gamma^0)$. We know that $Z_t^2\in K$ a. s. If A has rank $m>0$, then, for $t>0$, the support of $\mathcal{L}(Z_t^2)$ is an m-dimensional affine subspace of $\mathbb{R}^N$, which contradicts (ii) of Lemma 4.10. Hence $A=0$. It follows that $Z_t^2 = t\gamma^0$ and hence $\gamma^0\in K$. Thus all assertions in (c) are proved. ■

Definition 4.12 We call $\{Z_t\colon t\ge 0\}$ a *K-increasing Lévy process*, or *K-valued Lévy process*, or *K-valued subordinator*, if it satisfies the conditions in Theorem 4.11.

Example 4.13 $\mathbb{R}_+^N = [0,\infty)^N$ is a cone in $\mathbb{R}^N$. An $\mathbb{R}_+^N$-increasing Lévy process is sometimes called an *N-variate subordinator*.

For any $w = (w_j)_{1\le j\le N}$ and $v = (v_j)_{1\le j\le N}$ in $\mathbb{C}^N$, we define $\langle w, v\rangle = \sum_{j=1}^N w_j v_j$. This is not the Hermitian inner product.

Remark 4.14 Let $\{Z_t : t\ge 0\}$ be a K-valued subordinator. Then we have

$$E[e^{\langle w, Z_t\rangle}] = e^{t\Psi(w)} \tag{4.10}$$

with

$$\Psi(w) = \langle \gamma^0, w \rangle + \int_K (e^{\langle w,s \rangle} - 1)\nu(ds) \tag{4.11}$$

for any $w \in \mathbb{C}^N$ satisfying $\mathrm{Re}\,\langle w, s\rangle \le 0$ for all $s \in K$. If $\mathrm{Re}\,\langle w, s\rangle \le 0$ for $s \in K$, then $\left|e^{\langle w,s\rangle}\right| = e^{\mathrm{Re}\,\langle w,s\rangle} \le 1$ and both sides of (4.10) are definable. The equality is a special case of Theorem 25.17 of [93].

Example 4.15 Let $\{B_t^- : t \ge 0\}$ be a negative binomial subordinator with parameter $0 < p < 1$, that is, for $t > 0$,

$$P\left[B_t^- = n\right] = p^t \binom{n+t-1}{n} (1-p)^n, \qquad n = 0, 1, \ldots.$$

For each $j = 1, \ldots, N$ let $\{X_j(t) \colon t \ge 0\}$ be a Lévy process on $\mathbb{R}$ with $\mathcal{L}(X_j(t))$ being Γ-distribution with parameters λt and α ($\lambda > 0$ and $\alpha > 0$ do not depend on j):

$$P\left[X_j(t) \in B\right] = \int_{B\cap(0,\infty)} \frac{\alpha^{\lambda t}}{\Gamma(\lambda t)} x^{\lambda t-1} e^{-\alpha x} dx, \qquad B \in \mathcal{B}(\mathbb{R}).$$

Assume that $\{B_t^-\}$, $\{X_j(t)\}, \ldots, \{X_N(t)\}$ are independent. Define

$$Y_t = (Y_j(t))_{1\le j\le N} = (X_j(t + \lambda^{-1} B_{\lambda t}^-))_{1\le j\le N}.$$

Then $\{Y_t\}$ is an N-variate subordinator whose components are not independent. Each component $\{Y_j(t)\}$ is a Lévy process with $\mathcal{L}(Y_j(t))$ being Γ-distribution with parameters λt and $p\alpha$. Indeed,

$$\begin{aligned}
P\left[Y_j(t) \le u\right] &= P\left[X_j(t + \lambda^{-1} B_{\lambda t}^-) \le u\right] \\
&= \sum_{n=0}^{\infty} \int_0^u \frac{\alpha^{\lambda(t+\lambda^{-1}n)}}{\Gamma(\lambda(t+\lambda^{-1}n))} v^{\lambda(t+\lambda^{-1}n)-1} e^{-\alpha v} dv\, p^{\lambda t} \binom{n+\lambda t-1}{n} (1-p)^n \\
&= \int_0^u \left(\alpha^{\lambda t} v^{\lambda t-1} e^{-\alpha v} p^{\lambda t} \sum_{n=0}^{\infty} \frac{\alpha^n}{\Gamma(\lambda t+n)} v^n (1-p)^n \binom{n+\lambda t-1}{n} \right) dv
\end{aligned}$$

and, since $\binom{n+\lambda t-1}{n} = \Gamma(n+\lambda t)/(n!\,\Gamma(\lambda t))$, we get

$$P\left[Y_j(t) \le u\right] = \int_0^u \frac{(\alpha p)^{\lambda t}}{\Gamma(\lambda t)} v^{\lambda t-1} e^{-\alpha p v} dv.$$

If $N = 2$, then, for each $t > 0$, we can find the distribution of Y_t has density

$$C_t (y_1 y_2)^{(\lambda t-1)/2} e^{-\alpha(y_1+y_2)} I_{\lambda t-1}\left(2\alpha\sqrt{(1-p)y_1 y_2}\right)$$

on $\mathbb{R}_+^2$, where C_t is a positive constant depending on t and $I_{\lambda t-1}$ is the modified Bessel function of order $\lambda t - 1$ given by (4.8) of [93, p. 21].

4.2 Subordination of Cone-Parameter Lévy Processes

Cones are multidimensional analogues of $[0, \infty)$. We extend, in a natural way, the concept of a Lévy process to a process with parameter set being a cone. Further, we consider their associated cone-parameter convolution semigroups.

Let K be a cone in $\mathbb{R}^M$ in this section.

Definition 4.16 Let f be a mapping from K into $\mathbb{R}^d$.

(i) Let $s^0 \in K$. We say that f is *K-right continuous at* s^0, if, for every K-decreasing sequence $\{s^n\}_{n=1,2,\dots}$ in K with $|s^n - s^0| \to 0$, we have $|f(s^n) - f(s^0)| \to 0$. We say that f is *K-right continuous* if f is K-right continuous at every $s^0 \in K$.
(ii) Let $s^0 \in K \setminus \{0\}$. We say that f has *K-left limit at* s^0, if, for every K-increasing sequence $\{s^n\}_{n=1,2,\dots}$ in $K \setminus \{s^0\}$ satisfying $|s^n - s^0| \to 0$, $\lim_{n\to\infty} f(s^n)$ exists in $\mathbb{R}^d$. We say f has *K-left limits* if it has K-left limit at every $s^0 \in K \setminus \{0\}$.

Remark 4.17 We should keep in mind that, if f has K-left limit at s^0, $\lim_{n\to\infty} f(s^n)$ may depend on the choice of the sequence $\{s^n\}$. For example, if $K = \mathbb{R}_+^2$, $s^0 = (s_j^0)_{j=1,2} \in \mathbb{R}_+^2 \setminus \{0\}$, and $f(s) = f_1(s_1) + f_2(s_2)$ for $s = (s_j)_{j=1,2}$ and if, for each j, $f_j(s_j)$ is a step function with a jump at s_j^0, then, for an $\mathbb{R}_+^2$-increasing sequence $s^n = (s_j^n)_{j=1,2}$ in $\mathbb{R}_+^2 \setminus \{s^0\}$ satisfying $|s^n - s^0| \to 0$,

$$\lim_{n\to\infty} f(s^n) = \begin{cases} f_1(s_1^0-) + f_2(s_2^0-) & \text{if } s_j^n < s_j^0 \text{ for } j = 1, 2 \text{ for all } n, \\ f_1(s_1^0-) + f_2(s_2^0) & \text{if } s_1^n < s_1^0, s_2^n = s_2^0 \text{ for all } n, \\ f_1(s_1^0) + f_2(s_2^0-) & \text{if } s_1^n = s_1^0, s_2^n < s_2^0 \text{ for all } n. \end{cases}$$

Definition 4.18 A *K-parameter Lévy process* $\{X_s : s \in K\}$ on $\mathbb{R}^d$ is a collection of random variables on $\mathbb{R}^d$ satisfying the following conditions.

(a) If $n \geq 3$, $s^1, \dots, s^n \in K$, and $s^k \leq_K s^{k+1}$ for $k = 1, \dots, n-1$, then $X_{s^{k+1}} - X_{s^k}$, $k = 1, \dots, n-1$, are independent.
(b) If $s^1, \dots, s^4 \in K$ and $s^2 - s^1 = s^4 - s^3 \in K$, then $X_{s^2} - X_{s^1} \overset{d}{=} X_{s^4} - X_{s^3}$.
(c) For each $s \in K$, $X_{s'} \to X_s$ in probability as $|s' - s| \to 0$ with $s' \in K$.
(d) $X_0 = 0$ a. s.
(e) Almost surely, $X_s(\omega)$ is K-right continuous with K-left limits in s.

If $\{X_s : s \in K\}$ satisfies (a)–(d), then $\{X_s : s \in K\}$ is called a *K-parameter Lévy process in law* on $\mathbb{R}^d$.

Definition 4.19 A family $\{\mu_s : s \in K\}$ of distributions on $\mathbb{R}^d$ is called a *K-parameter convolution semigroup on $\mathbb{R}^d$* if

(i) $\mu_{s^1} * \mu_{s^2} = \mu_{s^1+s^2}$ for all $s^1, s^2 \in K$.
(ii) $\mu_{ts} \to \delta_0$ for $s \in K$ as $t \to 0+$ (that is, as $t > 0$ and $t \to 0$).

Remark 4.20 It follows from the property (i) of Definition 4.19 that $\mu_0 = \delta_0$. Indeed, we have $\mu_0 = \mu_0 * \mu_0$ and thus $\widehat{\mu}_0(z) = \widehat{\mu}_0(z)^2$, which implies $\widehat{\mu}_0(z) = 0$ or 1 for each z. We then have $\widehat{\mu}_0(z) = 1$ for all z, since $\widehat{\mu}_0(0) = 1$ and $\widehat{\mu}_0$ is continuous. Therefore $\mu_0 = \delta_0$. It also follows from (i) that $\mu_s \in ID(\mathbb{R}^d)$ for all $s \in K$ because, for each $n \in \mathbb{N}$, $n^{-1}s \in K$ and $\mu_s = (\mu_{n^{-1}s})^n$.

Lemma 4.21 *Let $\{X_s : s \in K\}$ be a K-parameter Lévy process on $\mathbb{R}^d$ and let $\mu_s = \mathcal{L}(X_s)$. Then the following are true.*

(i) $\{\mu_s : s \in K\}$ is a K-parameter convolution semigroup on $\mathbb{R}^d$.
(ii) $\{X_{ts^0} : t \geq 0\}$ is a Lévy process on $\mathbb{R}^d$ for any $s^0 \in K$.
(iii) μ_s is infinitely divisible for all $s \in K$.

Proof

(i) $\mathcal{L}(X_{s^1+s^2}) = \mathcal{L}((X_{s^1} - X_0) + (X_{s^1+s^2} - X_{s^1})) = \mathcal{L}(X_{s^1} - X_0) * \mathcal{L}(X_{s^1+s^2} - X_{s^1}) = \mathcal{L}(X_{s^1}) * \mathcal{L}(X_{s^2})$, since $X_{s^1} - X_0$ and $X_{s^1+s^2} - X_{s^1}$ are independent, $\mathcal{L}(X_{s^1+s^2} - X_{s^1}) = \mathcal{L}(X_{s^2} - X_0)$, and $X_0 = 0$ a. s. Since $X_{ts} \to X_0 = 0$ in probability as $t \downarrow 0$, $\mu_{ts} \to \delta_0$ as $t \downarrow 0$.
(ii) Fix $s^0 \in K$. If $0 \leq t_0 < \cdots < t_n$ with $n \geq 2$, then $X_{t_{j+1}s^0} - X_{t_j s^0}$, $j = 0, \ldots, n-1$, are independent. If $0 \leq s < t$, then $X_{ts^0} - X_{ss^0} \stackrel{d}{=} X_{(t-s)s^0}$. Note that $\lim_{t' \to t} P\left[\left|X_{t's^0} - X_{ts^0}\right| > \varepsilon\right] = 0$ and $X_{0s^0} = 0$ a.s. Finally, almost surely X_{ts^0} is right continuous with left limits in t from the property (e).
(iii) This follows from (i) by Remark 4.20. ∎

Remark 4.22 The $\{\mu_s : s \in K\}$ in Lemma 4.21 is called the *K-parameter convolution semigroup induced by the K-parameter Lévy process $\{X_s : s \in K\}$*. Conversely, is it true that any K-parameter convolution semigroup on $\mathbb{R}^d$ is induced by some K-parameter Lévy process on $\mathbb{R}^d$? The answer is negative, as will be shown in Theorem 4.48.

Now let us give an analogue of the first half of Theorem 4.3 on subordination.

Theorem 4.23 *Let $\{Z_t : t \geq 0\}$ be a K-valued subordinator and $\{X_s : s \in K\}$ a K-parameter Lévy process on $\mathbb{R}^d$. Suppose that they are independent. Define $Y_t = X_{Z_t}$. Then $\{Y_t : t \geq 0\}$ is a Lévy process on $\mathbb{R}^d$.*

Proof Let $n \geq 2$ and $f^1, \ldots, f^{n-1}$ be measurable and bounded from $\mathbb{R}^d$ to $\mathbb{R}$, and let $0 \leq t_1 \leq \cdots \leq t_n$. Let $s^k \in K$, $k = 1, \ldots, n$, with $s^1 \leq_K s^2 \leq_K \cdots \leq_K s^n$ and let

$$G\left(s^1, \ldots, s^n\right) = E\left[\prod_{k=1}^{n-1} f^k\left(X_{s^{k+1}} - X_{s^k}\right)\right].$$

Since $X_{s^{k+1}} - X_{s^k}$, $k = 1, \ldots, n-1$, are independent, we have

$$G\left(s^1, \ldots, s^n\right) = \prod_{k=1}^{n-1} E\left[f^k\left(X_{s^{k+1}} - X_{s^k}\right)\right].$$

Next, let $g^k(s) = E\left[f^k(X_s)\right]$ for $s \in K$. Since $X_{s^{k+1}} - X_{s^k} \stackrel{d}{=} X_{s^{k+1}-s^k}$, we have

$$E\left[f^k\left(X_{s^{k+1}} - X_{s^k}\right)\right] = g^k(s^{k+1} - s^k).$$

It follows that

$$G\left(s^1, \ldots, s^n\right) = \prod_{k=1}^{n-1} g^k\left(s^{k+1} - s^k\right).$$

We use the standard argument for independence (such as Proposition 1.16 of [93]). As $\{X_s\}$ and $\{Z_t\}$ are independent, we obtain

$$E\left[\prod_{k=1}^{n-1} f^k\left(Y_{t_{k+1}} - Y_{t_k}\right)\right] = E\left[G\left(Z_{t_1}, \ldots, Z_{t_n}\right)\right] = \prod_{k=1}^{n-1} E\left[g^k\left(Z_{t_{k+1}} - Z_{t_k}\right)\right], \tag{4.12}$$

noting that $Z_{t_1}, \ldots, Z_{t_n}$ make a K-increasing sequence. Choosing $f^j = 1$ for all $j \neq k$ and using that $\{Z_t\}$ has stationary increments, we see that

$$\begin{aligned} E\left[f^k\left(Y_{t_{k+1}} - Y_{t_k}\right)\right] &= E\left[g^k\left(Z_{t_{k+1}} - Z_{t_k}\right)\right] \\ &= E\left[g^k\left(Z_{t_{k+1}-t_k}\right)\right] = E\left[f^k\left(Y_{t_{k+1}-t_k}\right)\right]. \end{aligned} \tag{4.13}$$

Equations (4.12) and (4.13) say that $\{Y_t\}$ has independent increments and stationary increments, respectively. Evidently $Y_0 = 0$ a. s. Since Z_t is right continuous with left limits in t and since $\{X_s\}$ has property (e) in Definition 4.18, Y_t is right continuous with left limits in t a. s. Here notice that, when $t_n < t$ and $t_n \uparrow t$, we have $Z_{t_n} \leq_K Z_{t_{n+1}}$ and $Z_{t_n} \to Z_{t-}$, but Z_{t_n} can be equal to Z_{t-}; even if $Z_{t_n} = Z_{t-}$ for large n, $Y_{t_n} = X_{Z_{t_n}}$ is convergent. Now $\mathcal{L}(Y_t - Y_s) = \mathcal{L}(Y_{t-s}) \to \delta_0$ as $t \downarrow s$ or $s \uparrow t$, which shows stochastic continuity of $\{Y_t\}$. Thus $\{Y_t\}$ is a Lévy process on $\mathbb{R}^d$. ■

Definition 4.24 Let $M \geq 2$ and let K be a cone in $\mathbb{R}^M$. Let $\{Z_t : t \geq 0\}$ be a K-valued subordinator, $\{X_s : s \in K\}$ a K-parameter Lévy process on $\mathbb{R}^d$, and assume that they are independent. The procedure in Theorem 4.23 of making $\{Y_t\}$

from $\{X_s\}$ and $\{Z_t\}$ is called *multivariate subordination*. Here $\{X_s\}$, $\{Z_t\}$, and $\{Y_t\}$ are called, respectively, *subordinand*, *subordinating* (or *subordinator*), and *subordinated*.

The following lemma is useful in considering examples.

Lemma 4.25 *Let* $\{X_s^1 : s \in K\}, \ldots, \{X_s^n : s \in K\}$ *be independent* K*-parameter Lévy processes on* $\mathbb{R}^d$*. Let*

$$X_s = X_s^1 + \cdots + X_s^n.$$

Then $\{X_s : s \in K\}$ *is a* K*-parameter Lévy process on* $\mathbb{R}^d$*.*

Proof It is straightforward to check the defining properties for a K-parameter Lévy process. ■

Let us introduce the notions of strong basis and weak basis of a cone K in $\mathbb{R}^M$. The smallest linear subspace of $\mathbb{R}^M$ that contains K is called the *linear subspace generated by* K.

Definition 4.26 Let K be a cone in $\mathbb{R}^M$. If $\{e^1, \ldots, e^N\}$ is a linearly independent system in K such that $K = \{s_1 e^1 + \cdots + s_N e^N : s_1 \geq 0, \ldots, s_N \geq 0\}$, then $\{e^1, \ldots, e^N\}$ is called a *strong basis* of K. If $\{e^1, \ldots, e^N\}$ is a basis of the linear subspace generated by K and $e^1, \ldots, e^N$ are in K, then $\{e^1, \ldots, e^N\}$ is called a *weak basis* of K and K is called N*-dimensional cone*. We say that K is *nondegenerate* if it is M-dimensional.

Example 4.27 In one dimension, there are only two cones in $\mathbb{R}$; they are the intervals $[0, \infty)$ and $(-\infty, 0]$ and each of them has a strong basis. In two dimensions, let us identify $\mathbb{R}^2$ with the complex plane $\mathbb{C}$. Then, for any choice of $\theta_1 \in \mathbb{R}$ and $\theta_2 \in (\theta_1, \theta_1 + \pi)$, the sector $\{z = re^{i\theta} : r \geq 0 \text{ and } \theta \in [\theta_1, \theta_2]\}$ is a cone in $\mathbb{C}$ and it has a strong basis $\{e^{i\theta_1}, e^{i\theta_2}\}$. Any nondegenerate cone in $\mathbb{C}$ is of this form. In $\mathbb{R}^3$ any triangular cone is a nondegenerate cone having a strong basis, but any non-triangular nondegenerate cone does not have a strong basis. Any cone K in $\mathbb{R}^M$ has a weak basis. If $N = M$, then the nonnegative orthant $\mathbb{R}_+^N$ has the standard basis $\{e^1, \ldots, e^N\}$ of $\mathbb{R}^N$ as a strong basis.

Remark 4.28 Let K be a cone with strong basis $\{e^1, \ldots, e^N\}$. For $k = 1, 2$ let $s^k = s_1^k e^1 + \cdots + s_N^k e^N$ with $s_1^k, \ldots, s_N^k \in \mathbb{R}$. Then $s^1 \leq_K s^2$ if and only if $s_j^1 \leq s_j^2$ for $j = 1, \ldots, N$. Thus, in this sense, the partial order determined by K is equivalent to the componentwise order.

Definition 4.29 Let K be an N-dimensional cone in $\mathbb{R}^M$ and L the linear space generated by K. Let K' be a cone in $\mathbb{R}^{M'}$. We say that K and K' are *isomorphic* if there exists a linear transformation T from L to $\mathbb{R}^{M'}$ such that $\dim(TL) = N$ and $TK = K'$. We call T *isomorphism* and we denote by T^{-1} its inverse defined on TL.

From this definition we see that K' is an N-dimensional cone. Further, $u^1 \leq_{K'} u^2$ if and only if $T^{-1}u^1 \leq_K T^{-1}u^2$. A system $\{e^1, \ldots, e^N\}$ is a strong basis (resp. weak basis) of K' if and only if $\{T^{-1}e^1, \ldots, T^{-1}e^N\}$ is a strong basis (resp. weak basis) of K.

Proposition 4.30

(a) *Any N-dimensional cone K with a strong basis is isomorphic to $\mathbb{R}_+^N$.*
(b) *For any strong basis $\{e^1, \ldots, e^N\}$ of K there is an isomorphism given by (a) such that each $s_1e^1 + \cdots + s_Ne^N \in K$ is identified with $(s_j)_{1\leq j\leq N} \in \mathbb{R}_+^N$.*

Proof (a) Let $\{e^1, \ldots, e^N\}$ be a strong basis of K. Denote by $\{f^1, \ldots, f^N\}$ the standard basis of $\mathbb{R}^N$. Let us consider any bijective mapping from the strong basis $\{e^1, \ldots, e^N\}$ of K to the strong basis $\{f^1, \ldots, f^N\}$ of $\mathbb{R}_+^N$. It is readily seen that this mapping extends uniquely to a one-to-one linear transformation from the linear space generated by K onto $\mathbb{R}^N$.

Assertion (b) follows from (a) by considering the particular mapping $e^j \to f^j$, $j = 1, \ldots, N$. ■

Remark 4.31 If $\{e^1, \ldots, e^N\}$ and $\{f^1, \ldots, f^N\}$ are both strong bases of K, then they are identical up to scaling and permutation. See Proposition 2.4 of [71] for a proof.

4.3 Case of Cone $\mathbb{R}_+^N$

In this section we assume $K = \mathbb{R}_+^N$, a cone in $\mathbb{R}^N$. Denote the unit vectors $e^k = (\delta_{kj})_{1\leq j\leq N}$ for $k = 1, \ldots, N$, where $\delta_{kj} = 1$ or 0 according as $k = j$ or not. Thus $\{e^1, \ldots, e^N\}$ is the standard basis of $\mathbb{R}^N$ and, at the same time, the strong basis of the cone $\mathbb{R}_+^N$. The partial order $s^1 \leq_K s^2$ in $\mathbb{R}^N$ for $K = \mathbb{R}_+^N$ is equivalent to the componentwise order $s_j^1 \leq s_j^2$, $j = 1, \ldots, N$, for $s^k = (s_j^k)_{1\leq j\leq N} = s_1^k e^1 + \cdots + s_N^k e^N$, $k = 1, 2$. First we give various examples of $\mathbb{R}_+^N$-parameter Lévy processes and their relation to $\mathbb{R}_+^N$-parameter convolution semigroups. Then joint distributions of $\mathbb{R}_+^N$-parameter Lévy processes are considered. Generating triplets appearing in multivariate subordination are described.

Example 4.32 Let $\{V_t : t \geq 0\}$ be a Lévy process on $\mathbb{R}^d$. Fix $c = (c_j)_{1\leq j\leq N} \in \mathbb{R}_+^N$. Define

$$X_s = V_{\langle c,s\rangle} = V_{c_1s_1+\cdots+c_Ns_N} \quad \text{for } s = (s_j)_{1\leq j\leq N} \in \mathbb{R}_+^N. \tag{4.14}$$

Then, $\{X_s : s \in \mathbb{R}_+^N\}$ is a $\mathbb{R}_+^N$-parameter Lévy process on $\mathbb{R}^d$.

To show this, we write $K = \mathbb{R}_+^N$. If $s^1 \leq_K s^2 \leq_K \cdots \leq_K s^n$ with $s^1, \ldots, s^n \in K$, then $X_{s^{k+1}} - X_{s^k} = V_{\langle c,s^{k+1}\rangle} - V_{\langle c,s^k\rangle}$, $k = 1, \ldots, n$, are independent, since

$\langle c, s^{k+1}\rangle - \langle c, s^k\rangle \geq 0$. If $s^1 \leq_K s^2$ and $s^3 \leq_K s^4$ such that $s^2 - s^1 = s^4 - s^3$, then $X_{s^2} - X_{s^1} = V_{\langle c,s^2\rangle} - V_{\langle c,s^1\rangle} \stackrel{d}{=} V_{\langle c,s^2-s^1\rangle} = V_{\langle c,s^4-s^3\rangle} \stackrel{d}{=} X_{s^4} - X_{s^3}$. If $s' \in K$ and $s' \to s$, then $X_{s'} = V_{\langle c,s'\rangle} \to V_{\langle c,s\rangle} = X_s$ in probability. If $\{s^k\}_{k\geq 1}$ is a K-decreasing sequence converging to $s \in K$, then $|X_{s^k} - X_s| = \left|V_{\langle c,s^k-s\rangle}\right| \to 0$, since $\langle c, s^k - s\rangle \to 0$. If $\{s^k\}_{k\geq 1}$ is K-increasing, $s^k \neq s$, and $s^k \to s$, then $\langle c, s^k\rangle \leq \langle c, s^{k+1}\rangle$, $\langle c, s^k\rangle \leq \langle c, s\rangle$ and $\langle c, s^k\rangle \to \langle c, s\rangle$, and hence $X_{s^k} = V_{\langle c,s^k\rangle}$ is convergent to $V_{\langle c,s\rangle-}$ or $V_{\langle c,s\rangle}$. Thus X_s is K-right continuous with K-left limits a. s. Finally $X_0 = V_{\langle c,0\rangle} = 0$ a. s.

Example 4.33 Let $\{V_t^j : t \geq 0\}$, $j = 1, \ldots, N$, be independent Lévy processes on $\mathbb{R}^d$. Define

$$V_s = V_{s_1}^1 + V_{s_2}^2 + \cdots + V_{s_N}^N \quad \text{for } s = (s_j)_{1\leq j\leq N} \in \mathbb{R}_+^N. \tag{4.15}$$

Then $\{V_s : s \in \mathbb{R}_+^N\}$ is a $\mathbb{R}_+^N$-parameter Lévy process on $\mathbb{R}^d$.

Indeed, for each j, $\{V_{s_j}^j : s \in \mathbb{R}_+^N\}$ is a $\mathbb{R}_+^N$-parameter Lévy process, as it is a special case of Example 4.32 with $c = e^j$. Hence $\{V_s : s \in \mathbb{R}_+^N\}$ is a $\mathbb{R}_+^N$-parameter Lévy process by Lemma 4.25.

Example 4.34 For each $j = 1, \ldots, N$, let $\{U_t^j : t \geq 0\}$ be a Lévy process on $\mathbb{R}^{d_j}$. Assume that they are independent. Let $d = d_1 + \cdots + d_N$. Define

$$U_s = (U_{s_j}^j)_{1\leq j\leq N} \quad \text{for } s = (s_j)_{1\leq j\leq N} \in \mathbb{R}_+^N, \tag{4.16}$$

that is, U_s is the direct product of $U_{s_j}^j$, $j = 1, \ldots, N$. Then $\{U_s : s \in \mathbb{R}_+^N\}$ is a $\mathbb{R}_+^N$-parameter Lévy process on $\mathbb{R}^d$.

Indeed, for each k, let $\{X_s^k : s \in \mathbb{R}_+^N\}$ be the process defined as $X_s^k = (X_{s_j}^{k,j})_{1\leq j\leq N}$ for $s = (s_j)_{1\leq j\leq N}$ with $X_{s_j}^{k,j} = 0$ in $\mathbb{R}^{d_j}$ for $j \neq k$ and $X_{s_k}^{k,k} = U_{s_k}^k$. Then $\{X_s^k : s \in \mathbb{R}_+^N\}$ is a $\mathbb{R}_+^N$-parameter Lévy process on $\mathbb{R}^d$ just by the same proof as each term of (4.15) in Example 4.33. Then $\{X_s^k\}$, $k = 1, \ldots, N$, are independent, $U_s = X_s^1 + \cdots + X_s^N$, and Lemma 4.25 applies.

Example 4.35 Let $\{\mu_s : s \in \mathbb{R}_+^2\}$ be a collection of distributions on $\mathbb{R}$ such that, for each $s = (s_1, s_2)^\top$, μ_s is Gaussian distribution on $\mathbb{R}$ with mean zero and variance $s_1 + s_2$. Here $\top$ means *transpose* as in Notation. Then $\{\mu_s\}$ is a $\mathbb{R}_+^2$-parameter convolution semigroup. Let $\{V_t^j : t \geq 0\}$, $j = 1, 2, 3$, be independent Brownian motions on $\mathbb{R}$. Let $\{X_s^1 : s \in \mathbb{R}_+^2\}$ and $\{X_s^2 : s \in \mathbb{R}_+^2\}$ be defined, respectively, by $X_s^1 = V_{s_1}^1 + V_{s_2}^2$ and $X_s^2 = V_{s_1+s_2}^3$ for every $s = (s_1, s_2)^\top$. Then, clearly both $\{X_s^1\}$ and $\{X_s^2\}$ are $\mathbb{R}_+^2$-parameter Lévy processes and they induce a common $\mathbb{R}_+^2$-parameter convolution semigroup $\{\mu_s\}$, that is, $\mathcal{L}(X_s^1) = \mathcal{L}(X_s^2) = \mu_s$

for every $s \in \mathbb{R}^2_+$. Nevertheless, $\mathcal{L}\left(\left(X^1_{e^1}, X^1_{e^2}\right)^\top\right) \neq \mathcal{L}\left(\left(X^2_{e^1}, X^2_{e^2}\right)^\top\right)$, since $\mathcal{L}\left((V^1_1 + V^2_0, V^1_0 + V^2_1)^\top\right) = \mathcal{L}\left((V^1_1, V^2_1)^\top\right) \neq \mathcal{L}\left((V^3_1, V^3_1)^\top\right)$.

Joint distributions of cone-parameter Lévy processes are not necessarily infinitely divisible as is shown by the following.

Example 4.36 Let $\{\mu_s : s \in \mathbb{R}^2_+\}$, $\{X^1_s : s \in \mathbb{R}^2_+\}$, $\{X^2_s : s \in \mathbb{R}^2_+\}$, and e^1, e^2 be as in Example 4.35. Let $X_s = X^1_{Us} + X^2_{(1-U)s}$, where U is a Bernoulli random variable with $0 < p = P(U = 1) < 1$ and $P(U = 0) = 1 - p$ such that U is independent of $\{\{X^1_s\}, \{X^2_s\}\}$. Then $\{X_s : s \in \mathbb{R}^2_+\}$ is a $\mathbb{R}^2_+$-parameter Lévy process which induces the $\mathbb{R}^2_+$-parameter convolution semigroup $\{\mu_s\}$ by Proposition 3.9 of [72]. The joint distribution $\mathcal{L}((X_{e^1}, X_{e^2})^\top)$ of $\{X_s\}$ is not infinitely divisible by Remark 3.11 of [72].

Theorem 4.37 *Let us write $\mathbb{R}^N_+ = K$. Let $\{X_s : s \in K\}$ be a K-parameter Lévy process on $\mathbb{R}^d$. Define $X^j_t = X_{te^j}$ and let $\{V^j_t : t \geq 0\}$, $j = 1, \ldots, N$, be independent Lévy processes on $\mathbb{R}^d$ such that $\{V^j_t\} \stackrel{d}{=} \{X^j_t\}$ for each j. Define $\{V_s : s \in K\}$ as*

$$V_{s_1e^1+\cdots+s_Ne^N} = V^1_{s_1} + \cdots + V^N_{s_N}. \tag{4.17}$$

Then, for every $n \in \mathbb{N}$ and $s^1, \ldots, s^n \in K$ satisfying

$$s^1 \leq_K s^2 \leq_K \cdots \leq_K s^n, \tag{4.18}$$

we have

$$(X_{s^k})_{1\leq k\leq n} \stackrel{d}{=} (V_{s^k})_{1\leq k\leq n}. \tag{4.19}$$

Proof We claim that

$$X_s \stackrel{d}{=} V_s \quad \text{for } s \in K. \tag{4.20}$$

Indeed, for $s = (s_j)_{1\leq j\leq N} = s_1e^1 + \cdots + s_Ne^N \in K$,

$$X_s = X_{s_1e^1} + \left(X_{s_1e^1+s_2e^2} - X_{s_1e^1}\right) + \cdots + \left(X_s - X_{s_1e^1+\cdots+s_{N-1}e^{N-1}}\right).$$

The right-hand side is the sum of N independent terms by the definition of K-parameter Lévy process. Further,

$$X_{s_1e^1} = X^1_{s_1} \stackrel{d}{=} V^1_{s_1},$$

$$X_{s_1e^1+s_2e^2} - X_{s_1e^1} \stackrel{d}{=} X_{s_2e^2} = X^2_{s_2} \stackrel{d}{=} V^2_{s_2},$$

and so on. Hence we obtain (4.20) from (4.17). Now we claim that (4.19) holds for $s^1, \ldots, s^n \in K$ satisfying (4.18). In order to prove this, it is enough to prove

$$\left(X_{s^k} - X_{s^{k-1}}\right)_{1 \le k \le n} \stackrel{d}{=} \left(V_{s^k} - V_{s^{k-1}}\right)_{1 \le k \le n}, \tag{4.21}$$

where $s^0 = 0$, since there is an $n \times n$ matrix T such that

$$\left(X_{s^k}\right)_{1 \le k \le n} = T\left(\left(X_{s^k} - X_{s^{k-1}}\right)_{1 \le k \le n}\right),$$
$$\left(V_{s^k}\right)_{1 \le k \le n} = T\left(\left(V_{s^k} - V_{s^{k-1}}\right)_{1 \le k \le n}\right).$$

Since $\{V_s\}$ also is a K-parameter Lévy process by Example 4.33, the components of each side of (4.21) are independent. Furthermore,

$$X_{s^k} - X_{s^{k-1}} \stackrel{d}{=} X_{s^k - s^{k-1}} \stackrel{d}{=} V_{s^k - s^{k-1}} \stackrel{d}{=} V_{s^k} - V_{s^{k-1}}$$

by virtue of (4.20). Hence we have (4.21). Therefore (4.19) holds. ■

Corollary 4.38 *Let $\{X_s : s \in \mathbb{R}_+^N\}$ be a $\mathbb{R}_+^N$-parameter Lévy process on $\mathbb{R}^d$ and define $X_t^j = X_{te^j}$. Then*

$$Ee^{i\langle z, X_s \rangle} = \prod_{j=1}^{N} Ee^{i\langle z, X_{s_j}^j \rangle} \quad \textit{for } s = (s_j)_{1 \le j \le N},\ z \in \mathbb{R}^d. \tag{4.22}$$

Proof This is an expression of (4.19) when $n = 1$. ■

Remark 4.39 In the notation of Theorem 4.37 the joint distribution $\mathcal{L}\left((X_{s^k})_{1 \le k \le n}\right)$ of a K-parameter Lévy process $\{X_s : s \in K\}$ is determined by $\mathcal{L}\left(X_{e^j}\right)$, $j = 1, \ldots, N$, as long as (4.18) is satisfied. In particular, for each s, $\mathcal{L}\left(X_s\right)$ is determined by $\mathcal{L}\left(X_{e^j}\right)$, $j = 1, \ldots, N$. However, general joint distributions are not determined by $\mathcal{L}\left(X_{e^j}\right)$, $j = 1, \ldots, N$. For example, suppose that $X_s = W_{s_1 + \cdots + s_N}$ for $s = (s_j)_{1 \le j \le N} \in K$ with a Lévy process $\{W_t : t \ge 0\}$ as in Example 4.32 with $c_j = 1$, $j = 1, \ldots, N$. Then $X_{e^1} = X_{e^2} = \cdots = X_{e^N}$ while $V_{e^1}, V_{e^2}, \ldots, V_{e^N}$ are independent. Thus $\mathcal{L}\left(\left(X_{e^j}\right)_{1 \le j \le N}\right) \ne \mathcal{L}\left(\left(V_{e^j}\right)_{1 \le j \le N}\right)$ except in the trivial case, although $\mathcal{L}\left(X_{e^j}\right) = \mathcal{L}\left(V_{e^j}\right)$, $j = 1, \ldots, N$.

Let us extend Lemma 4.5 to a $\mathbb{R}_+^N$-parameter Lévy process.

Lemma 4.40 *Let $\{X_s : s \in \mathbb{R}_+^N\}$ be a $\mathbb{R}_+^N$-parameter Lévy process on $\mathbb{R}^d$. Then, there are constants $C(\varepsilon)$, C_1, C_2, C_3 such that*

$$P[|X_s| > \varepsilon] \le C(\varepsilon)|s| \quad \textit{for } \varepsilon > 0, \tag{4.23}$$

$$E[|X_s|^2; |X_s| \le 1] \le C_1|s|, \tag{4.24}$$

$$|E[X_s; |X_s| \le 1]| \le C_2|s|, \tag{4.25}$$

$$E[|X_s|; |X_s| \le 1] \le C_3|s|^{1/2}. \tag{4.26}$$

Proof We use Theorem 4.37. Since $X_s \overset{d}{=} V_s$, it is enough to show the estimates for V_s. Notice that $\sum_{j=1}^N |s_j| \le C|s|$ for some constant $C > 0$. Proof of (4.23) and (4.24) uses Lemma 4.5 for $V_{s_j}^j$ as follows:

$$\begin{aligned} P\left[|V_s| > \varepsilon\right] &= P\left[\left|\sum_{j=1}^N V_{s_j}^j\right| > \varepsilon\right] \le P\left[|V_{s_j}^j| > \varepsilon/N \text{ for some } j\right] \\ &\le \sum_{j=1}^N P\left[|V_{s_j}^j| > \varepsilon/N\right] \le \widetilde{C}(\varepsilon)\sum_{j=1}^N s_j, \end{aligned}$$

for some constant $\widetilde{C}(\varepsilon)$.

$$\begin{aligned} E\left[|V_s|^2; |V_s| \le 1\right] &\le E\left[\left|\sum_{j=1}^N V_{s_j}^j\right|^2; |V_{s_j}^j| \le 1 \;\forall j\right] + P\left[|V_{s_j}^j| > 1 \text{ for some } j\right] \\ &\le N^2 \sum_{j=1}^N E\left[|V_{s_j}^j|^2; |V_{s_j}^j| \le 1\right] + \sum_{j=1}^N P\left[|V_{s_j}^j| > 1\right] \\ &\le \widetilde{C}_1 \sum_{j=1}^N s_j, \end{aligned}$$

for some constant $\widetilde{C}_1$.

We denote the kth component by putting the superscript (k) in order to prove (4.25). We have

$$|E\left[V_s; |V_s| \le 1\right]| \le \sum_{k=1}^N \left|E\left[iV_s^{(k)}; |V_s| \le 1\right]\right| = \sum_{k=1}^N |I_{k1} + I_{k2} + I_{k3}|,$$

where

$$\begin{aligned} I_{k1} &= E\left[e^{iV_s^{(k)}} - 1\right], \qquad I_{k2} = -E\left[e^{iV_s^{(k)}} - 1; |V_s| > 1\right], \\ I_{k3} &= -E\left[e^{iV_s^{(k)}} - 1 - iV_s^{(k)}; |V_s| \le 1\right]. \end{aligned}$$

We have

$$\begin{aligned} I_{k1} &= E\left[e^{i\sum_{j=1}^N V_{s_j}^{j(k)}} - 1\right] \\ &= E\left[e^{i\sum_{j=1}^N V_{s_j}^{j(k)}} - e^{i\sum_{j=1}^{N-1} V_{s_j}^{j(k)}}\right] + \cdots + E\left[e^{iV_{s_1}^{1(k)}} - 1\right] \end{aligned}$$

and hence

$$|I_{k1}| \le \sum_{j=1}^N \left| E\left[e^{iV_{s_j}^{j(k)}} - 1 \right] \right| = \sum_{j=1}^N \left| \left(E\left[e^{iV_1^{j(k)}} \right] \right)^{s_j} - 1 \right| \le \widetilde{C}_2 \sum_{j=1}^N s_j,$$

for some constant $\widetilde{C}_2$. As we have

$$|I_{k2}| \le 2P\left[|V_s| > 1\right] \le 2C(1)|s|,$$
$$|I_{k3}| \le \tfrac{1}{2} E\left[(V_s^{(k)})^2; |V_s| \le 1 \right] \le \tfrac{1}{2} E\left[|V_s|^2; |V_s| \le 1 \right] \le \tfrac{1}{2} C_1 |s|$$

by (4.23) and (4.24), we now obtain (4.25). Finally

$$E\left[|V_s|; |V_s| \le 1\right] \le \left(E\left[|V_s|^2; |V_s| \le 1 \right] \right)^{1/2} \le C_1{}^{1/2} |s|^{1/2}$$

by Schwarz's inequality. ■

Let us give description of generating triplets in multivariate subordination in the case of the cone $\mathbb{R}_+^N$.

Theorem 4.41 *Let $\{Y_t : t \ge 0\}$ be a Lévy process on $\mathbb{R}^d$ obtained by multivariate subordination from an $\mathbb{R}_+^N$-parameter Lévy process $\{X_s : s \in \mathbb{R}_+^N\}$ on $\mathbb{R}^d$ and $\mathbb{R}_+^N$-valued subordinator $\{Z_t : t \ge 0\}$ as in Theorem 4.23. Let $X_t^j = X_{te^j}$.*

(i) The characteristic function of Y_t is as follows:

$$Ee^{i\langle z, Y_t\rangle} = e^{t\Psi_Z(\psi_X(z))}, \quad z \in \mathbb{R}^d, \tag{4.27}$$

where Ψ_Z is the function Ψ of (4.11) in Remark 4.14 and

$$\psi_X(z) = (\psi_X^j(z))_{1 \le j \le N}, \qquad \psi_X^j(z) = \left(\log \widehat{\mathcal{L}(X_1^j)} \right)(z). \tag{4.28}$$

(ii) Let ν_Z and $\gamma_Z^0 = (\gamma_{Z,j}^0)_{1 \le j \le N}$ be the Lévy measure and the drift of $\{Z_t\}$ and let $(A_X^j, \nu_X^j, \gamma_X^j)$ be the generating triplet of $\{X_t^j\}$. Let $\mu_s = \mathcal{L}(X_s)$. Then the generating triplet (A_Y, ν_Y, γ_Y) of $\{Y_t\}$ is as follows:

$$A_Y = \sum_{j=1}^N \gamma_{Z,j}^0 A_X^j, \tag{4.29}$$

$$\nu_Y(B) = \int_{\mathbb{R}_+^N} \mu_s(B) \nu_Z(ds) + \sum_{j=1}^N \gamma_{Z,j}^0 \nu_X^j(B), \quad B \in \mathcal{B}(\mathbb{R}^d \setminus \{0\}), \tag{4.30}$$

$$\gamma_Y = \int_{\mathbb{R}_+^N} \nu_Z(ds) \int_{|x| \le 1} x \mu_s(dx) + \sum_{j=1}^N \gamma_{Z,j}^0 \gamma_X^j. \tag{4.31}$$

(iii) If $\int_{|s|\le 1}|s|^{1/2}\nu_Z(ds)<\infty$ and $\gamma_Z^0=0$, then $A_Y=0$, $\int_{|x|\le 1}|x|\nu_Y(dx)<\infty$, and the drift γ_Y^0 of $\{Y_t\}$ is zero.

Proof

(i) Let $\left\{V_t^j : t\ge 0\right\}$, $j=1,2,\ldots,N$, and $\{V_s : s\in\mathbb{R}_+^N\}$ be the processes defined in Theorem 4.37. Then, by (4.22) of Corollary 4.38,

$$Ee^{i\langle z,X_s\rangle}=\prod_{j=1}^{N}Ee^{i\langle z,X^j_{s_j}\rangle}=\prod_{j=1}^{N}e^{s_j\psi_X^j(z)}=e^{\langle s,\psi_X(z)\rangle} \tag{4.32}$$

for $z\in\mathbb{R}^d$ and $s\in\mathbb{R}_+^N$. Use the standard argument for independence (as in Proposition 1.16 of [93]). We get

$$Ee^{i\langle z,Y_t\rangle}=E\left[\left(Ee^{i\langle z,X_s\rangle}\right)_{s=Z_t}\right]=Ee^{\langle Z_t,\psi_X(z)\rangle}=e^{t\Psi_Z(\psi_X(z))}$$

for $z\in\mathbb{R}^d$ by (4.10), since $\mathrm{Re}\,\langle\psi_X(z),s\rangle=\sum_{j=1}^{N}(\mathrm{Re}\,\psi_X^j(z))s_j\le 0$. This is (4.27).

(ii) Let $z\in\mathbb{R}^d$. Let $K=\mathbb{R}_+^N$. We have

$$Ee^{i\langle z,Y_t\rangle}=e^{t\Psi_Z(\psi_X(z))}=\exp\left[t\left(\langle\gamma_Z^0,\psi_X(z)\rangle+\int_K(e^{\langle\psi_X(z),s\rangle}-1)\nu_Z(ds)\right)\right]$$

by (4.11), since $\mathrm{Re}\,\langle\psi_X(z),s\rangle\le 0$. Notice that

$$\begin{aligned}\langle\gamma_Z^0,\psi_X(z)\rangle&=\sum_{j=1}^{N}\gamma_{Z,j}^0\psi_X^j(z)\\&=\sum_{j=1}^{N}\gamma_{Z,j}^0\left(-\frac{1}{2}\langle z,A_X^jz\rangle+i\langle\gamma_X^j,z\rangle+\int_{\mathbb{R}^d}g(z,x)\nu_X^j(dx)\right)\end{aligned}$$

with $g(z,x)=e^{i\langle z,x\rangle}-1-i\langle z,x\rangle 1_{\{|x|\le 1\}}(x)$. Hence

$$\begin{aligned}\langle\gamma_Z^0,\psi_X(z)\rangle=&-\frac{1}{2}\left\langle z,\sum_{j=1}^{N}\gamma_{Z,j}^0A_X^jz\right\rangle+i\left\langle\sum_{j=1}^{N}\gamma_{Z,j}^0\gamma_X^j,z\right\rangle\\&+\int_{\mathbb{R}^d}g(z,x)\left(\sum_{j=1}^{N}\gamma_{Z,j}^0\nu_X^j\right)(dx).\end{aligned}$$

Next it follows from (4.32) that

$$\int_K (e^{\langle \psi_X(z),s\rangle} - 1)\nu_Z(ds) = \int_K (Ee^{i\langle z,X_s\rangle} - 1)\nu_Z(ds)$$

$$= \int_K \nu_Z(ds) \int_{\mathbb{R}^d} (e^{i\langle z,x\rangle} - 1)\mu_s(dx)$$

$$= \int_K \nu_Z(ds) \int_{\mathbb{R}^d} g(z,x)\mu_s(dx) + i \int_K \nu_Z(ds) \left\langle z, \int_{|x|\le 1} x\mu_s(dx) \right\rangle.$$

Here we used (4.25) of Lemma 4.40 below and $\int_{|s|\le 1} |s|\nu_Z(ds) < \infty$. Define $\widetilde{\nu}$ by $\widetilde{\nu}(B) = \int_K \mu_s(B \setminus \{0\})\nu_Z(ds)$, $B \in \mathcal{B}(\mathbb{R}^d)$. Then, by (4.23) and (4.24) of Lemma 4.40,

$$\int_{|x|\le 1} |x|^2 \widetilde{\nu}(dx) \le C_1 \int_K |s|\nu_Z(ds) < \infty,$$

$$\int_{|x|>1} \widetilde{\nu}(dx) \le C(1) \int_K |s|\nu_Z(ds) < \infty.$$

Hence

$$\int_K (e^{\langle \psi_X(z),s\rangle} - 1)\nu_Z(ds) = \int_{\mathbb{R}^d} g(z,x)\widetilde{\nu}(dx) + i \left\langle \int_K \nu_Z(ds) \int_{|x|\le 1} x\mu_s(dx), z \right\rangle.$$

Thus we get (4.29)–(4.31).

(iii) Assume $\int_{|s|\le 1} |s|^{1/2}\nu_Z(ds) < \infty$ and $\gamma_Z^0 = 0$. Then $A_Y = 0$ by (4.29),

$$\int_{|x|\le 1} |x|\nu_Y(dx) = \int_K \nu_Z(ds) \int_{|x|\le 1} |x|\mu_s(dx) < \infty$$

by (4.26) of Lemma 4.40, and

$$\gamma_Y^0 = \gamma_Y - \int_{|x|\le 1} x\nu_Y(dx) = \int_K \nu_Z(ds) \int_{|x|\le 1} x\mu_s(dx) - \int_{|x|\le 1} x\nu_Y(dx) = 0$$

by (4.30) and (4.31). ■

Remark 4.42 Theorem 4.41 shows that the distribution of $\{Y_t\}$ subordinate to $\{X_s\}$ by $\{Z_t\}$ is determined by the distributions of $\{X_t^1\}, \ldots, \{X_t^N\}$, and $\{Z_t\}$, although the joint distributions of $\{X_s\}$ are not in general determined by $\{X_t^1\}, \ldots, \{X_t^N\}$ as Remark 4.39 says. This is because relevant joint distributions of $\{X_s\}$ are only those with $\mathbb{R}_+^N$-increasing sequences of parameters and they are determined by $\{X_t^1\}, \ldots, \{X_t^N\}$ as in Theorem 4.37.

4.4 Case of General Cone K

In order to make clear the relation between K-parameter Lévy processes and K-parameter convolution semigroups for a cone K, we introduce the notions of generativeness and related ones for K-parameter convolution semigroups.

Definition 4.43 Let K be a cone in $\mathbb{R}^M$. Let $\{\mu_s : s \in K\}$ be a K-parameter convolution semigroup. A K-parameter Lévy process in law $\{X_s : s \in K\}$ is said to be *associated* with $\{\mu_s : s \in K\}$ if $\mathcal{L}(X_s) = \mu_s$. We say that $\{\mu_s : s \in K\}$ is *generative* if there is a K-parameter Lévy process in law associated with it; otherwise, $\{\mu_s : s \in K\}$ is *non-generative.* We say that $\{\mu_s : s \in K\}$ is *unique-generative* if it is generative and any two K-parameter Lévy processes in law $\{X_s^1 : s \in K\}$ and $\{X_s^2 : s \in K\}$ associated with it satisfy $\{X_s^1 : s \in K\} \stackrel{d}{=} \{X_s^2 : s \in K\}$. If $\{\mu_s : s \in K\}$ is generative but not unique-generative, we say that it is *multiple-generative.*

Remark 4.44 Existence of multiple-generative K-parameter convolution semigroups on $\mathbb{R}^d$ follows from Example 4.35 and Remark 4.39.

In the case of cone K with a strong basis, main results on generativeness and unique-generativeness of K-parameter convolution semigroups are contained in the following theorem. We call a subset L of $\mathbb{R}^d$ an *additive subgroup* of $\mathbb{R}^d$ if $x - y \in L$ for all $x, y \in L$.

Theorem 4.45 *Let K be a cone in $\mathbb{R}^M$ with a strong basis $\{e^1, \dots, e^N\}$ and let $\{\mu_s : s \in K\}$ be a K-parameter convolution semigroup on $\mathbb{R}^d$. Let $V_s = V_{s_1}^1 + \cdots + V_{s_N}^N$ for $s = s_1 e^1 + \cdots + s_N e^N$ in K, where $\left\{V_t^j : t \geq 0\right\}$, $j = 1, \dots, N$, are independent Lévy processes satisfying $\mathcal{L}(V_1^j) = \mu_{e^j}$ for $j = 1, \dots, N$. Then*

(i) *$\{\mu_s : s \in K\}$ is generative. In particular, $\{V_s : s \in K\}$ is a K-parameter Lévy process associated with $\{\mu_s\}$.*
(ii) *The following three statements are equivalent:*
   (a) *$\{\mu_s : s \in K\}$ is unique-generative.*
   (b) *Any K-parameter Lévy process in law $\{X_s : s \in K\}$ associated with $\{\mu_s : s \in K\}$ satisfies $\{X_s : s \in K\} \stackrel{d}{=} \{V_s : s \in K\}$.*
   (c) *For any K-parameter Lévy process in law $\{X_s : s \in K\}$ associated with $\{\mu_s : s \in K\}$ we have $X_s = X_{s_1 e^1} + \cdots + X_{s_N e^N}$ almost surely for $s = s_1 e^1 + \cdots + s_N e^N \in K$.*

(iii) *For $j = 1, \dots, N$ let L_j be an additive subgroup of $\mathbb{R}^d$ such that $L_j \in \mathcal{B}(\mathbb{R}^d)$. Assume that $L_j \cap L_k = \{0\}$ for all distinct j, k. If $\mu_{te^j}(L_j) = 1$ for all $t \geq 0$ for $j = 1, \dots, N$, then $\{\mu_s\}$ is unique-generative.*

Proof of (i) is similar to Theorem 4.37 for $K = \mathbb{R}_+^N$. We can also prove (ii) in the same idea. Proof of (iii) is rather technical; see Theorem 4.2 of [72].

Example 4.46 Let $K = \mathbb{R}_+^2$ and let $\{e^1, e^2\}$ be the standard basis of $\mathbb{R}^2$. Let $L_1 = \mathbb{Q}^d$ and $L_2 = (c\mathbb{Q})^d$ with $c \in \mathbb{R} \setminus \mathbb{Q}$. Let $\{\mu_s : \mathbb{R}_+^2\}$ be the compound Poisson convolution semigroup on $\mathbb{R}^d$ such that, for each j, the Lévy measure of μ_{e^j} is concentrated in L_j. Then $\{\mu_s\}$ unique-generative, as Theorem 4.45 (iii) applies. This is Example 4.4 of [72].

The following definition and two subsequent theorems are concerned with existence of non-generative cone-parameter convolution semigroups.

Definition 4.47 Let $d \geq 2$ and let S_d^+ be the set of symmetric nonnegative-definite $d \times d$ matrices. Each $s = (s_{jk})_{j,k=1}^d \in S_d^+$ is determined by the lower triangle $(s_{jk})_{j \leq k}$ with $d(d+1)/2$ entries. The set S_d^+ is identified with a subset of $\mathbb{R}^{d(d+1)/2}$. Then S_d^+ is a nondegenerate cone in $\mathbb{R}^{d(d+1)/2}$. For $s \in S_d^+$ let μ_s be Gaussian distribution on $\mathbb{R}^d$ with mean zero and covariance matrix s. Then, obviously, $\{\mu_s : s \in S_d^+\}$ is an S_d^+-parameter convolution semigroup on $\mathbb{R}^d$. It is called the *canonical* S_d^+*-parameter convolution semigroup.*

Theorem 4.48 *The canonical* S_d^+*-parameter convolution semigroup on* $\mathbb{R}^d$ *is non-generative.*

This can be rephrased as follows: in the sense of cone-parameter Lévy process, there is no Brownian motion on $\mathbb{R}^d$ with parameter in S_d^+. Theorem 4.48 is a consequence of the following more general theorem. We say that a K-parameter convolution semigroup $\{\mu_s : s \in K\}$ is *trivial* if μ_s is trivial for all $s \in K$.

Theorem 4.49 *Let* $d \geq 2$ *and let* $K = S_d^+$. *Let* $\{\mu_s : s \in K\}$ *be a non-trivial* K*-parameter convolution semigroup on* $\mathbb{R}^d$ *such that* $\int |x|^2 \mu_s(dx) < \infty$ *and the covariance matrix* v_s *of* μ_s *satisfies* $v_s \leq_K s$ *for all* $s \in K$. *Then* $\{\mu_s : s \in K\}$ *is non-generative.*

This is by Pedersen and Sato [72, Theorem 4.1] (2004). In their proof assertion (iii) on unique-generativeness in Theorem 4.45 is effectively used.

Remark 4.50 The cone S_d^+ has no strong basis. A proof of it is given by Theorem 4.48 combined with Theorem 4.45.

Some sufficient conditions for generativeness of K-parameter convolution semigroups for general cone K are known.

Theorem 4.51 *Let* K *be a cone in* $\mathbb{R}^M$ *and let* $\{\mu_s : s \in K\}$ *be a* K*-parameter convolution semigroup on* $\mathbb{R}^d$. *Then the following two statements are true.*

(i) If $d = 1$, *then* $\{\mu_s : s \in K\}$ *is generative.*
(ii) If μ_s *is purely non-Gaussian for all* $s \in K$, *then* $\{\mu_s : s \in K\}$ *is generative.*

See Theorems 5.2 and 5.3 of [72].

We have shown in Theorem 4.23 that given a K-parameter Lévy process and a K-valued subordinator for a cone K, we can perform multivariate subordination. However, Theorem 4.41 shows that, at least for $K = \mathbb{R}_+^N$, the distribution of the subordinated Lévy process is determined by the induced K-parameter convolution semigroup and the distribution of the K-valued subordinator. This suggests that it is more natural to study subordination of K-parameter convolution semigroups than to study subordination of K-parameter Lévy processes. This study is made by Pedersen and Sato [71] (2003) and the description of the generating triplets of the subordinated convolution semigroup is found for a general cone K. Next we recall these results. First, we give some basic properties of K-parameter convolution semigroups.

Proposition 4.52 *Let K be a cone in $\mathbb{R}^M$. Let $\{e^1, \dots, e^N\}$ be a weak basis of K. Then any K-parameter convolution semigroup $\{\mu_s : s \in K\}$ on $\mathbb{R}^d$ has the following properties.*

(i) $\mu_s \in ID(\mathbb{R}^d)$ *for each* $s \in K$, *and* $\mu_{ts} = \mu_s^{t*}$ *for* $t \geq 0$.

(ii) For $s = s_1 e^1 + \cdots + s_N e^N \in K$ *we have*

$$\widehat{\mu}_s(z) = (\widehat{\mu}_{e^1}^{s_1})(z) \cdots (\widehat{\mu}_{e^N}^{s_N})(z), \qquad z \in \mathbb{R}^d, \tag{4.33}$$

where $(\widehat{\mu}_{e^j}^{s_j})(z) = \exp\{s_j (\log \widehat{\mu}_{e^j})(z)\}$ *as defined in Definition 1.3.*

(iii) Let (A_s, ν_s, γ_s) *be the generating triplet of* μ_s. *For* $s = s_1 e^1 + \cdots + s_N e^N \in K$ *we have*

$$A_s = s_1 A_{e^1} + \cdots + s_N A_{e^N},$$

$$\nu_s = s_1 \nu_{e^1} + \cdots + s_N \nu_{e^N},$$

$$\gamma_s = s_1 \gamma_{e^1} + \cdots + s_N \gamma_{e^N}.$$

(iv) For any sequence $\{s^n\}_{n=1,2,\dots}$ *with* $|s^n - s^0| \to 0$ *as* $n \to \infty$, *we have* $\mu_{s^n} \to \mu_{s^0}$ *as* $n \to \infty$.

Recall that some of $s_1, \dots, s_n$ may be negative unless $\{e^1, \dots, e^N\}$ is a strong basis.

Proof

(i) For each $n \in \mathbb{N}$, $\mu_s = (\mu_{(1/n)s})^{n*}$. Hence $\mu_s \in ID(\mathbb{R}^d)$ and $\mu_{(1/n)s} = \mu_s^{(1/n)*}$. Thus $\mu_{(m/n)s} = \mu_s^{(m/n)*}$. Since $\widehat{\mu}_{ts}$ is right continuous in t by the property (ii) of Definition 4.19 we have $\widehat{\mu}_{ts} = \widehat{\mu}_s^t$.

(ii) Write $s_j = s_j^+ - s_j^-$, where $s_j^+ = s_j \vee 0$ and $s_j^- = -(s_j \wedge 0)$. We have $s = u - v$ where $u = s_1^+ e^1 + \cdots + s_N^+ e^N \in K$ and $v = s_1^- e^1 + \cdots + s_N^- e^N \in K$. Then $\mu_s * \mu_v = \mu_u$. Hence, by (4.33),

$$\widehat{\mu}_s(z) = \frac{\widehat{\mu}_u(z)}{\widehat{\mu}_v(z)} = \frac{\left(\widehat{\mu}_{e^1}^{s_1^+}\right)(z) \cdots \left(\widehat{\mu}_{e^N}^{s_N^+}\right)(z)}{\left(\widehat{\mu}_{e^1}^{s_1^-}\right)(z) \cdots \left(\widehat{\mu}_{e^N}^{s_N^-}\right)(z)},$$

which is (4.33).

(iii) Follows from (4.33) by use of Exercise 12.2 of [93].
(iv) also follows from (4.33), since $|s^n - s^0| \to 0$ implies $s_j^n \to s_j^0$ for $j = 1,\dots,N$. ■

Let us proceed to generalization of subordination. For any measure μ on $\mathbb{R}^d$ and any μ-integrable function f, we denote $\mu(f) = \int_{\mathbb{R}^d} f(x)\mu(dx)$.

Theorem 4.53 *Let K_1 be an N_1-dimensional cone in $\mathbb{R}^{M_1}$ and K_2 an N_2-dimensional cone in $\mathbb{R}^{M_2}$. Let $\{\mu_u : u \in K_2\}$ be a K_2-parameter convolution semigroup on $\mathbb{R}^d$ and $\{\rho_s : s \in K_1\}$ a K_1-parameter convolution semigroup on $\mathbb{R}^{M_2}$ such that $\rho_s(K_2) = 1$ for all $s \in K_1$. Define a distribution σ_s on $\mathbb{R}^d$ by*

$$\sigma_s(f) = \int_{K_2} \mu_u(f)\rho_s(du). \tag{4.34}$$

for all bounded continuous functions f on $\mathbb{R}^d$. Then the family $\{\sigma_s : s \in K_1\}$ is a K_1-parameter convolution semigroup on $\mathbb{R}^d$.

See Theorem 4.3 of Pedersen and Sato [71] (2003). Notice that $\mu_u(f)$ in (4.34) is continuous in u by virtue of Proposition 4.52 (iv).

Definition 4.54 The above procedure for obtaining $\{\sigma_s : s \in K_1\}$ is called *subordination* of $\{\mu_u : u \in K_2\}$ by $\{\rho_s : s \in K_1\}$. The convolution semigroups $\{\mu_u : u \in K_2\}$, $\{\rho_s : s \in K_1\}$, and $\{\sigma_s : s \in K_1\}$ are, respectively, called *subordinand*, *subordinating* (or *subordinator*), and *subordinated*.

The following theorem reduces to Theorem 4.3 in the case of ordinary subordination $K_1 = K_2 = \mathbb{R}_+$. In the case of multivariate subordination $K_1 = \mathbb{R}_+$ and $K_2 = K$ it reduces to Theorem 4.41.

Theorem 4.55 *Let $\{\mu_u : u \in K_2\}$, $\{\rho_s : s \in K_1\}$, and $\{\sigma_s : s \in K_1\}$ be the subordinand, subordinating, and subordinated convolution semigroups in Theorem 4.53. Let $\{h^1,\dots,h^{N_2}\}$ be a weak basis of K_2. Let $\left(A_\mu^j, \nu_\mu^j, \gamma_\mu^j\right)$ be the generating triplet of μ_{h^j} for $j = 1,\dots,N_2$. Let ν_{ρ_s} and $\gamma_{\rho_s}^0$ be the Lévy measure and the drift of ρ_s for all $s \in K_1$ and let us decompose*

$$\gamma_{\rho_s}^0 = \gamma_{\rho_s,1}^0 h^1 + \cdots + \gamma_{\rho_s,N_2}^0 h^{N_2}.$$

Let R be the orthogonal projection from $\mathbb{R}^{M_2}$ to the linear subspace L_2 generated by K_2 and let T be a linear transformation from $\mathbb{R}^{M_2}$ onto $\mathbb{R}^{N_2}$ defined by $Tu = (u_j)_{1\le j\le N_2}$ where $Ru = u_1h^1 + \cdots + u_{N_2}h^{N_2}$. Then

(i) For any $s \in K_1$, the characteristic function of σ_s is as follows:

$$\int_{\mathbb{R}^d} e^{i\langle z,x\rangle}\sigma_s(dx) = e^{\Psi_s^\rho(w)}, \qquad z \in \mathbb{R}^d,$$

where

$$\Psi_s^{\rho}(w) = \langle T\gamma_{\rho_s}^0, w\rangle + \int_{K_2} (e^{\langle w, Tu\rangle} - 1)\nu_{\rho_s}(du),$$

$$w = (w_j)_{1\le j\le N_2}, \quad w_j = -\tfrac{1}{2}\left\langle z, A_\mu^j z\right\rangle + \int_{\mathbb{R}^d} g(z,x)\nu_\mu^j(dx) + i\left\langle \gamma_\mu^j, z\right\rangle$$

with $g(z,x)$ *of (1.19).*

(ii) For any $s \in K_1$*, the generating triplet* $(A_{\sigma_s}, \nu_{\sigma_s}, \gamma_{\sigma_s})$ *of* σ_s *is as follows:*

$$A_{\sigma_s} = \sum_{j=1}^{N_2} \gamma_{\rho_s,j}^0 A_\mu^j,$$

$$\nu_{\sigma_s}(B) = \int_{K_2} \mu_u(B)\nu_{\rho_s}(du) + \sum_{j=1}^{N_2} \gamma_{\rho_s,j}^0 \nu_\mu^j(B), \quad B \in \mathcal{B}(\mathbb{R}^d \setminus \{0\}),$$

$$\gamma_{\sigma_s} = \int_{K_2} \nu_{\rho_s}(du)\int_{|x|\le 1} x\mu_u(dx) + \sum_{j=1}^{N_2} \gamma_{\rho_s,j}^0 \gamma_\mu^j.$$

(iii) Fix $s \in K_1$*. If* $\int_{K_2\cap\{|u|\le 1\}} |u|^{1/2}\nu_{\rho_s}(du) < \infty$ *and* $\gamma_{\rho_s}^0 = 0$*, then* $A_{\sigma_s} = 0$*,* $\int_{|x|\le 1} |x|\nu_{\sigma_s}(dx) < \infty$*, and the drift* $\gamma_{\sigma_s}^0$ *of* σ_s *is zero.*

This is Theorem 4.4 of Pedersen and Sato [71] (2003). Recall that some $\gamma_{\rho_s,j}^0$ may be negative unless $\{h^1, \dots, h^{N_2}\}$ is a strong basis of K_2.

We can give the description of the generating triplet of the subordinated in the multivariate subordination of Definition 4.24 based on Theorem 4.23 for general K as a special case of Theorem 4.55 with $K_1 = [0, \infty)$.

Notes

The idea of subordination was originated by Bochner and expounded in his book [14] (1955). Cone-parameter Lévy processes and their subordination have been studied in [8, 72, 74] (2001, 2004, 2005).

When $K = \mathbb{R}_+^N$, K-parameter Lévy processes and their subordination were introduced in [8] (2001). In this case of $K = \mathbb{R}_+^N$, all results in this chapter on multivariate subordination are found in [8] (2001). But the proof of Theorem 4.41 has been simplified. Example 4.15 is from [8] (2001); several other examples of construction of N-variate subordinators are also contained in that paper. In the case of a general cone K, all results in this chapter on K-parameter convolution semigroups are found in [71, 72] (2003, 2004); Theorem 4.23 is from there.

Furthermore, Examples 4.35 and 4.36, Definition 4.47, and related results are from [72] (2004).

In Sect. 4.2 the notion of subordination has been extended to the case of parameters in a general cone K in $\mathbb{R}^N$. We mention that Bochner [14] (1955) already considered processes with parameter in a cone, under the name of multidimensional time variable. The cone-parameter Lévy processes and their subordination studied in [8] (2001) and [72] (2004) are intimately connected with cone-parameter convolution semigroups and their subordination introduced by Pedersen and Sato [71] (2003).

The description of the generating triplet of the subordinated process in multivariate subordination of $\mathbb{R}_+^N$-parameter Lévy process is made in [8] (2001). See Theorem 4.41. More generally, in the case of subordination of cone-parameter convolution semigroups this description is made for all cones K in $\mathbb{R}^N$ in [71] (2003). See Theorem 4.55. Furthermore, since any cone in $\mathbb{R}^N$ with strong basis is isomorphic with $\mathbb{R}_+^N$, the results of [8] (2001) apply to any cone in $\mathbb{R}^N$ with strong basis. See Proposition 4.30.

Subordination and description of the generating triplet of the subordinated process have also been studied in the infinite-dimensional case. The paper [74] (2005) deals with subordination of cone-parameter Lévy processes with values in the Banach space of trace-class operators of a separable Hilbert space by a subordinator with values in the Banach space of trace-class operators. This paper describes the generating triplet of the subordinated process similarly to (4.29)–(4.31). Pérez-Abreu and Rocha-Arteaga [73] (2003) studied subordination of a Banach-space-valued Lévy process by a real-valued subordinator. They clarified the generating triplet of the subordinated Banach-space-valued Lévy process.

In the Gaussian case the multiparameter Brownian motion $\{B_s : s \in \mathbb{R}^N\}$ and the Brownian sheet $\{W_s : s \in \mathbb{R}^N\}$ have been discussed for a long time. We mention Lévy [55] (1948), Chentsov [18] (1957), and McKean [66] (1963) for the former and Orey and Pruitt [70] (1973) and Khoshnevisan and Shi [47] (1999) for the latter. When the parameter s is restricted to a cone K not isomorphic to $[0, \infty)$, neither $\{B_s : s \in K\}$ nor $\{W_s : s \in K\}$ is a K-parameter Lévy process. Likewise, two-parameter Lévy processes in Vares [131] (1983) and Lagaize [52] (2001) are not K-parameter Lévy processes in our sense. But probabilistic potential theory for the $\mathbb{R}_+^N$-parameter Lévy process in Example 4.33 with $\{V_t^j\}$, $j = 1, \ldots, N$, being a symmetric Lévy processes was studied by Hirsch [34] (1995) and, in the case where $\{V_t^j\}$ was a Brownian motion on $\mathbb{R}^d$ for each j, Khoshnevisan and Shi [47] (1999) called it the (N, d) additive Brownian motion and studied its capacity.

Theorem 4.11 in Sect. 4.1 on Lévy processes taking values in a cone is by Skorohod [115] (1991). In the terminology of Theorem 4.11, for a cone K in $\mathbb{R}^N$, K-increasing Lévy process on $\mathbb{R}^N$ is determined by an infinitely divisible distribution concentrated on K. An infinitely divisible distribution μ on $\mathbb{R}^N$ with generating triplet (A, ν, γ) is concentrated on K if and only if

$$A = 0, \ \nu\left(\mathbb{R}^N \setminus K\right) = 0, \ \int_{|x| \le 1} |x| \, \nu(dx) < \infty, \ \text{and } \gamma^0 \in K \tag{4.35}$$

where $\gamma^0 = \gamma - \int_{|x|\le 1} x\nu(dx)$, the drift of μ. Furthermore, in this case, μ has the Laplace transform $L_\mu(u)$, defined for u in the *dual cone* $K' = \{u \in \mathbb{R}^N : \langle u, x\rangle \ge 0 \text{ for all } x \in K\}$, given by

$$L_\mu(u) = \exp\left\{-\left\langle \gamma^0, u\right\rangle + \int_K \left(e^{-\langle u,x\rangle} - 1\right)\nu(dx)\right\}. \tag{4.36}$$

The existence of such representations for cone-increasing Lévy processes on infinite-dimensional spaces is not guaranteed in general. They are intrinsically related to functional-analytic aspects, specifically to the geometry of the cone. In fact, this result has been extended to infinite dimensions as follows.

Let B denote a separable Banach space with norm $\|\cdot\|$. Let $\mathcal{B}(B)$ be the class of Borel sets in B. Measures and distributions on B are understood to be defined on $\mathcal{B}(B)$. Terms such as B-valued random variable, its distribution, independence, convolution, infinite divisibility, cone, and K-increasing sequence are defined in the same way as in the case of $\mathbb{R}^N$. Denote by B' the strong topological dual of B. For any cone K we define the *dual cone* $K' = \{f \in B' : f(s) \ge 0 \text{ for all } s \in K\}$. Let μ be a distribution on B. The mapping $\widehat{\mu} : B' \to \mathbb{C}$ defined as $\widehat{\mu}(f) = \int_B e^{if(x)}\mu(dx)$ for $f \in B'$ is called the *characteristic functional* of μ. The mapping $L_\mu : K' \to \mathbb{R}$ defined as $L_\mu(f) = \int_B e^{-f(x)}\mu(dx)$ for $f \in K'$ is called the *Laplace transform* of μ.

With any infinitely divisible distribution μ on B, a triplet of parameters (A, ν, γ) from the Lévy–Khintchine representation of $\widehat{\mu}$ is associated. Here A is related to the Gaussian part of μ, ν is a σ-finite measure on $B\setminus\{0\}$ such that the mapping $f \longmapsto \exp\{\int_B \left(e^{if(x)} - 1 - if(x)1_{\{\|x\|\le 1\}}(x)\right)\nu(dx)\}$ for all $f \in B'$, is the characteristic functional of some distribution on B, and $\gamma \in B$; see Araujo and Giné [3, Theorem 6.3.2] (1991) and Linde [56, Theorem 5.7.3] (1986). The triplet (A, ν, γ) is unique and it is called the *generating triplet* of μ; and ν is called the *Lévy measure* of μ.

A sequence $\{s_n\}_{n=1,2,\ldots}$ in a cone K is called *K-majorized* if there exists $s \in K$ such that $s_n \le_K s$ for $n \ge 1$. A cone K is said to be *regular* if every K-increasing and K-majorized sequence in K is norm convergent. A cone K is called *normal* if for each $y \in K$ there is a constant $\lambda > 0$ such that $0 \le_K x \le_K y$ implies $\|x\| \le \lambda \|y\|$. Let c_0 be the Banach space of real sequences converging to zero with the supremum norm. Let c_0^+ be the cone in c_0 of nonnegative sequences converging to zero.

The following result is interesting in comparison with Theorem 4.11.

Let K be a normal cone in a separable Banach space B. Then, the following three statements are equivalent.

(i) The cone K is regular.
(ii) Every infinitely divisible distribution μ concentrated on K has characteristic functional

$$\widehat{\mu}(f) = \exp\left\{if(\gamma^0) + \int_K \left(e^{if(x)} - 1\right)\nu(dx)\right\}, \quad f \in B', \tag{4.37}$$

where $\gamma^0 \in K$ and ν is the Lévy measure concentrated on K satisfying $\nu(\{0\}) = 0$ and $\int_K (1 \wedge |f(x)|)\nu(dx) < \infty$ for $f \in B'$.

(iii) Every infinitely divisible distribution μ concentrated on K has Laplace transform

$$L_\mu(f) = \exp\left\{-f(\gamma^0) + \int_K \left(e^{-f(x)} - 1\right) \nu(dx)\right\}, \quad f \in K', \tag{4.38}$$

where $\gamma^0 \in K$ and ν is the Lévy measure concentrated on K satisfying $\nu(\{0\}) = 0$ and $\int_K (1 \wedge f(x))\nu(dx) < \infty$ for $f \in K'$.

This is by Pérez-Abreu and Rosiński [77, Theorem 1 and Remark 2] (2007). They also found that c_0 contains two normal cones such that one is regular and the other is not. Thus, the existence of representations (4.37) and (4.38) indeed depends on the cone. The two representations generalize (4.35) and (4.36), respectively, to Banach spaces. Dettweiler [22] (1976) showed (4.38) for cone-valued infinitely divisible random variables when K is a normal regular cone in a locally convex toplogical vector space.

A stochastic process $\{X_t : t \geq 0\}$ on B is called *Lévy process* if it has independent and stationary increments, is stochastically continuous with respect to the norm, and starts at 0 almost surely and if, almost surely, $X_t(\omega)$ is right continuous with left limits in t with respect to the norm. For any B-valued Lévy process $\{X_t : t \geq 0\}$ the distribution $\mathcal{L}(X_1)$ is infinitely divisible and the generating triplet (A, ν, γ) of $\mathcal{L}(X_1)$ is called the *generating triplet* of $\{X_t\}$; and ν is called the *Lévy measure* of $\{X_t\}$.

Henceforth let K be a cone in a separable Banach space B.

A process is a K-increasing Lévy process on B if and only if it is a K-valued Lévy process. The proof of this fact is like that of the equivalence between (a) and (b) of Theorem 4.11. In general separable Banach spaces the equivalence between (c) and (a)–(b) of Theorem 4.11 is not true.

We say that a K-increasing Lévy process on B is a *K-valued subordinator* or simply *K-subordinator*.

The existence of representation (4.37) for a K-subordinator was given (for different subclasses of cones) by Gihman and Skorohod [25] (1975), Pérez-Abreu and Rocha-Arteaga [74, 75] (2005, 2006), and Rocha-Arteaga [81] (2006). A K-subordinator $\{Z_t : t \geq 0\}$ is called *regular K-subordinator* if $\mu = \mathcal{L}(Z_1)$ has representation (4.37). Theorem 18 of [75] (2006) shows that the concepts of K-subordinator and regular K-subordinator coincide for a wide subclass of normal regular cones (normal cones not isomorphic with c_0^+). However, there are regular K-subordinators of unbounded variation; this is the case when $\int_{\|x\|\leq 1} x\nu(dx)$ is a Pettis integral but not a Bochner integral, [75, p. 48] (2006). The paper [81] (2006) shows that, for a class of Banach spaces, the representations (4.37) and (4.38) are completely analogous to (4.35) and (4.36) for a cone in $\mathbb{R}^N$ and regular subordinators share many properties with real subordinators, for instance the bounded variation property. Also, [81] (2006) contains examples of construction of subordinators in a class of cones with strong bases.

Representation (4.37) in the case of cone-valued additive processes on the duals of nuclear Fréchet spaces was proved by Pérez-Abreu et al. [76] (2005).

Barndorff-Nielsen and Pérez-Abreu [9] (2008) studied matrix subordinators with values in the cone S_d^+ of $d \times d$ symmetric nonnegative-definite matrices. In particular, they introduced and characterized a class of infinitely divisible distributions on the open subcone of positive definite-matrices of S_d^+ ([9] Theorem 4.3). This class is an analogue of the Goldie–Steutel–Bondesson class of infinitely divisible distributions on $\mathbb{R}_+$ studied by Bondesson [15] (1981). Concerning the Goldie–Steutel–Bondesson class of distributions on $\mathbb{R}^d$, see Notes in Chap. 2.

Γ-distributions on $\mathbb{R}_+$ are infinitely divisible. Pérez-Abreu and Stelzer [78] (2014) introduced Γ-distributions on a cone in $\mathbb{R}^N$ and studied the class of generalized Γ-convolutions on a cone. Also examples of nonnegative-definite matrix Γ-distributions are introduced.

Chapter 5
Inheritance in Multivariate Subordination

We now study inheritance of L_m property or strict stability from subordinator to subordinated in multivariate subordination. In order to observe this inheritance, we have to assume strict stability of the distribution at each $s \in K$ of a K-parameter subordinand $\{X_s : s \in K\}$. Section 5.1 gives results and examples. Section 5.2 discusses some generalization where the defining condition of selfdecomposability or stability for distributions on $\mathbb{R}^d$ involves a $d \times d$ matrix Q. This is called operator generalization.

5.1 Inheritance of L_m Property and Strict Stability

We begin with the following theorem and examples in the usual subordination.

Theorem 5.1 *Suppose that $\{X_t : t \geq 0\}$ is a strictly α-stable process on $\mathbb{R}^d$, $\{Z_t : t \geq 0\}$ is a subordinator, and they are independent. Let $\{Y_t : t \geq 0\}$ be a Lévy process on $\mathbb{R}^d$ constructed from $\{X_t\}$ by subordination by $\{Z_t\}$.*

(i) If $\{Z_t\}$ is selfdecomposable, then $\{Y_t\}$ is selfdecomposable.
(ii) More generally, let $m \in \{0, 1, \ldots, \infty\}$. If $\{Z_t\}$ is of class $L_m(\mathbb{R})$, then $\{Y_t\}$ is of class $L_m(\mathbb{R}^d)$.
(iii) If $\{Z_t\}$ is strictly β-stable, then $\{Y_t\}$ is strictly $\alpha\beta$-stable.

Halgreen [30] (1979) and Ismail and Kelker [36] (1979) proved part of these results. Proof of Theorem 5.1 will be given as a special case of Theorem 5.9.

Example 5.2 Let $0 < \alpha < 1$. Let $\{Y_t\}$ be the Lévy process on $\mathbb{R}$ subordinate to a strictly α-stable increasing process $\{X_t\}$ on $\mathbb{R}$ with $Ee^{-uX_t} = e^{-tu^\alpha}$, $u \geq 0$, by a Γ-process $\{Z_t\}$ with $EZ_1 = 1$. Then

$$P[Y_1 \leq x] = 1 - E_\alpha(-x^\alpha), \quad x \geq 0,$$

A. Rocha-Arteaga, K. Sato, *Topics in Infinitely Divisible Distributions and Lévy Processes*, SpringerBriefs in Probability and Mathematical Statistics,
https://doi.org/10.1007/978-3-030-22700-5_5

where $E_\alpha(x)$ is the Mittag–Leffler function $E_\alpha(x) = \sum_{n=0}^{\infty} x^n / \Gamma(n\alpha + 1)$, and

$$P[Y_t \le x] = \sum_{n=0}^{\infty} \frac{(-1)^n \Gamma(t+n)}{n!\,\Gamma(t)\,\Gamma(1+\alpha(t+n))} x^{\alpha(t+n)}, \quad x \ge 0.$$

By Theorem 5.1, $\mathcal{L}(Y_t)$ is selfdecomposable. See Pillai [79] (1990) or Sato [93, E 34.4] (1999).

Example 5.3 Let $0 < \alpha \le 2$. Let $\{Y_t\}$ be the Lévy process subordinate to a symmetric α-stable process $\{X_t\}$ on $\mathbb{R}$ with $Ee^{izX_t} = e^{-t|z|^\alpha}$ by a Γ-process $\{Z_t\}$ with $EZ_1 = 1/q$, $q > 0$. Then

$$Ee^{izY_t} = (1 + q^{-1}|z|^\alpha)^{-t}, \quad z \in \mathbb{R},$$

where $\mathcal{L}(Y_1)$ is Linnik distribution or geometric stable distribution (Example 4.7). Theorem 5.1 shows that $\mathcal{L}(Y_t)$ is selfdecomposable.

In the definitions and examples below, we use γ, δ, λ, χ, and ψ for parameters of some special distributions, keeping a customary usage.

Definition 5.4 The distribution

$$\mu_{\gamma,\delta}(dx) = (2\pi)^{-1/2}\delta e^{\gamma\delta} x^{-3/2} e^{-(\delta^2 x^{-1} + \gamma^2 x)/2} 1_{(0,\infty)}(x) dx$$

with parameters $\gamma > 0$, $\delta > 0$ is called *inverse Gaussian distribution.*

The Laplace transform $L_{\mu_{\gamma,\delta}}(u)$, $u \ge 0$, of $\mu_{\gamma,\delta}$ is

$$\begin{aligned}
L_{\mu_{\gamma,\delta}}(u) &= \int_{(0,\infty)} e^{-ux} \mu_{\gamma,\delta}(dx) = \exp\left[-\delta\left(\sqrt{2u+\gamma^2} - \gamma\right)\right] \\
&= \exp\left[2^{-1}\pi^{-1/2}\delta \int_0^\infty \left(e^{-(2u+\gamma^2)x} - 1\right) x^{-3/2} dx + \gamma\delta\right] \\
&= \exp\left[2^{-1}\pi^{-1/2}\delta \int_0^\infty \left(e^{-2ux} - 1\right) x^{-3/2} e^{-\gamma^2 x} dx\right] \\
&= \exp\left[(2\pi)^{-1/2}\,\delta \int_0^\infty \left(e^{-ux} - 1\right) x^{-3/2} e^{-\gamma^2 x/2} dx\right].
\end{aligned}$$

The last formula shows that $\mu_{\gamma,\delta}$ is infinitely divisible with Lévy measure density

$$(2\pi)^{-1/2}\delta x^{-3/2} e^{-\gamma^2 x/2}$$

on $(0, \infty)$. Hence $\mu_{\gamma,\delta}$ is selfdecomposable by Theorem 1.34.

For every $\lambda \in \mathbb{R}$ we denote by K_λ the *modified Bessel function of order* λ given by (4.9), (4.10) of [93, p. 21] .

Example 5.5 Let $\{Z_t\}$ be a subordinator with $\mathcal{L}(Z_1)$ being the inverse Gaussian $\mu_{\gamma,\delta}$. Then $\mathcal{L}(Z_t) = \mu_{\gamma,t\delta}$. Let $\{Y_t\}$ be the Lévy process subordinate to Brownian motion $\{X_t\}$ on $\mathbb{R}$ by $\{Z_t\}$. Then

$$\begin{aligned} P[Y_t \in B] &= \int_0^\infty \mu_{\gamma,t\delta}(ds) \int_B (2\pi s)^{-1/2} e^{-x^2/(2s)} dx \\ &= (2\pi)^{-1} t\delta e^{t\gamma\delta} \int_B dx \int_0^\infty s^{-2} e^{-((x^2+t^2\delta^2)/(2s))-(\gamma^2 s/2)} ds \\ &= (4\pi)^{-1} t\gamma^2\delta e^{t\gamma\delta} \int_B dx \int_0^\infty u^{-2} e^{-(\gamma^2(x^2+t^2\delta^2)/(4u))-u} du \\ &= \int_B \frac{\gamma e^{t\gamma\delta}}{\pi\sqrt{1+(x/(t\gamma))^2}} K_1\left(t\gamma\delta\sqrt{1+(x/(t\gamma))^2}\right) dx, \end{aligned}$$

where K_1 is the modified Bessel function of order 1. This shows that $\mathcal{L}(Y_t)$ is a special case of the *normal inverse Gaussian distribution* defined by Barndorff-Nielsen [5] (1997). By Theorem 5.1, it is selfdecomposable. By Theorem 4.3 its characteristic function is

$$Ee^{izY_t} = e^{t\Psi(-z^2/2)} = \exp\left[-t\delta\left(\sqrt{z^2+\gamma^2} - \gamma\right)\right]$$

with $\Psi(w) = -\delta\left(\sqrt{-2w+\gamma^2} - \gamma\right)$.

Definition 5.6 The distribution

$$\mu_{\lambda,\chi,\psi}(dx) = c x^{\lambda-1} e^{-(\chi x^{-1}+\psi x)/2} 1_{(0,\infty)}(x) dx$$

is called *generalized inverse Gaussian distribution* with parameters λ, χ, ψ. Here c is a normalizing constant. The domain of the parameters is given by $\{\lambda < 0,\ \chi > 0,\ \psi \geq 0\}$, $\{\lambda = 0,\ \chi > 0,\ \psi > 0\}$, and $\{\lambda > 0,\ \chi \geq 0,\ \psi > 0\}$.

The Laplace transform $L_{\mu_{\lambda,\chi,\psi}}(u)$, $u \geq 0$, of $\mu_{\lambda,\chi,\psi}$ is

$$L_{\mu_{\lambda,\chi,\psi}}(u) = \begin{cases} \left(\dfrac{\psi}{\psi+2u}\right)^{\lambda/2} \dfrac{K_\lambda\left(\sqrt{\chi(\psi+2u)}\right)}{K_\lambda\left(\sqrt{\chi\psi}\right)} & \text{if } \chi > 0 \text{ and } \psi > 0 \\ \dfrac{2^{1+(\lambda/2)} K_\lambda\left(\sqrt{2\chi u}\right)}{\Gamma(-\lambda)(\chi u)^{\lambda/2}} & \text{if } \lambda < 0,\ \chi > 0, \text{ and } \psi = 0, \end{cases}$$

where K_λ is the modified Bessel function of order λ. It is known that $\mu_{\lambda,\chi,\psi}$ is infinitely divisible and, moreover, selfdecomposable [93, E. 34.13]. It belongs to the smaller class GGC called *generalized Γ-convolutions*, which means that it is the limit of a sequence of convolutions of Γ-distributions. See Halgreen [30] (1979). Concerning this class, see Notes at end of Chap. 2.

In order to extend Theorem 5.1 to multivariate subordination, we prepare two lemmas.

Lemma 5.7 *Let K be a cone in $\mathbb{R}^N$ and $\{X_s : s \in K\}$ a K-parameter Lévy process on $\mathbb{R}^d$. Let $0 < \alpha \le 2$. Then $\mathcal{L}(X_s) \in \mathfrak{S}_\alpha^0$ if and only if $X_{ts} \stackrel{d}{=} t^{1/\alpha} X_s$ for every $t > 0$.*

Proof Let $\mu_s = \mathcal{L}(X_s)$. The meaning of $\mu_s \in \mathfrak{S}_\alpha^0$ is that $\mu_s \in ID$ and $\widehat{\mu}_s(z)^t = \widehat{\mu}_s(t^{1/\alpha} z)$ for $t > 0$. See Definition 1.23 and Proposition 1.22. Since, by Lemma 4.21, $\{X_{ts} : t \ge 0\}$ is a Lévy process, $\widehat{\mu}_{ts}(z) = \widehat{\mu}_s(z)^t$. Hence the condition is written as $X_{ts} \stackrel{d}{=} t^{1/\alpha} X_s$. ∎

Lemma 5.8 *Let $\{Z_t\}$ be a K-valued subordinator such that $\mathcal{L}(Z_t) \in L_0(\mathbb{R}^N)$ for $t \ge 0$. Let $\Psi(w)$ be the function in (4.11). For $b > 1$ define $\Psi_b(w)$ as*

$$\Psi(w) = \Psi(b^{-1} w) + \Psi_b(w). \tag{5.1}$$

Then $e^{t\Psi_b(iz)}$, $z \in \mathbb{R}^N$, is the characteristic function of a K-valued subordinator $\{Z_t^{(b)}\}$. Let $m \ge 1$. Then $\mathcal{L}(Z_t) \in L_m$ for $t \ge 0$ if and only if $\mathcal{L}(Z_t^{(b)}) \in L_{m-1}$ for $t \ge 0$ and $b > 1$.

Proof Let $\mu = \mathcal{L}(Z_1)$ with generating triplet (A, ν, γ). Its characteristic function is $\widehat{\mu}(z) = e^{\Psi(iz)}$, $z \in \mathbb{R}^N$. If $b > 1$, then by selfdecomposability there is a distribution ρ_b such that

$$\widehat{\mu}(z) = \widehat{\mu}\left(b^{-1} z\right) \widehat{\rho}_b(z).$$

Let μ_b be such that $\widehat{\mu}_b(z) = \widehat{\mu}\left(b^{-1} z\right)$. Then $\mu = \mu_b * \rho_b$ and, by Proposition 1.13, μ_b and ρ_b are in ID. Let $\left(\widetilde{A}_b, \widetilde{\nu}_b, \widetilde{\gamma}_b\right)$ and (A_b, ν_b, γ_b) be the generating triplets of μ_b and ρ_b, respectively. Then $A = \widetilde{A}_b + A_b$, $\nu = \widetilde{\nu}_b + \nu_b$, and $\gamma = \widetilde{\gamma}_b + \gamma_b$. Hence $\nu_b \le \nu$. By Theorem 4.11, $A = 0$, $\nu\left(\mathbb{R}^N \setminus K\right) = 0$, $\int_{|s|\le 1} |s|\, \nu(ds) < \infty$, and $\gamma^0 \in K$. Therefore $\nu_b\left(\mathbb{R}^N \setminus K\right) = 0$, and $\int_{|s|\le 1} |s|\, \nu_b(ds) < \infty$. Also $A_b = 0$, as $0 \le \langle z, A_b z\rangle \le \langle z, Az\rangle = 0$. Further, their drifts are related as $\gamma^0 = \widetilde{\gamma}_b^0 + \gamma_b^0$ and $\widetilde{\gamma}_b^0 = b^{-1}\gamma^0$. Thus $\gamma_b^0 = \left(1 - b^{-1}\right)\gamma^0 \in K$. Then, by Theorem 4.11, a Lévy process $\{Z_t^{(b)}\}$ with $\mathcal{L}(Z_1^{(b)}) = \rho_b$ is a K-valued subordinator. Its characteristic function equals $(\widehat{\rho}_b^t)(z) = e^{t\Psi_b(iz)}$. Finally, $\mathcal{L}(Z_t)$ is of class L_m if and only if, for each $b > 1$, $\rho_b \in L_{m-1}$, that is, $\mathcal{L}(Z_t^{(b)})$ is of class L_{m-1}. ∎

Theorem 5.9 *Let K be a cone in $\mathbb{R}^N$ and $0 < \alpha \le 2$. Let $\{Z_t : t \ge 0\}$ be a K-valued subordinator and $\{X_s : s \in K\}$ a K-parameter Lévy process on $\mathbb{R}^d$ such that $\mathcal{L}(X_s) \in \mathfrak{S}_\alpha^0$ for all $s \in K$. Assume that they are independent. Let $\{Y_t : t \ge 0\}$ be the Lévy process on $\mathbb{R}^d$ constructed from $\{X_s\}$ and $\{Z_t\}$ by multivariate subordination of Definition 4.24.*

(i) If $\{Z_t\}$ is selfdecomposable, then $\{Y_t\}$ is selfdecomposable.
(ii) Let $m \in \{0, 1, \ldots, \infty\}$. If $\{Z_t\}$ is of class $L_m(\mathbb{R}^N)$, then $\{Y_t\}$ is of class $L_m(\mathbb{R}^d)$.
(iii) Let $0 < \beta \le 1$. If $\mathcal{L}(Z_t) \in \mathfrak{S}^0_\beta$ for all $t \ge 0$, then $\mathcal{L}(Y_t) \in \mathfrak{S}^0_{\alpha\beta}$ for all $t \ge 0$.

Proof Let $\mu_s = \mathcal{L}(X_s)$.

(i) Let $\{Z_t\}$ be selfdecomposable. Then $\mathcal{L}(Z_t) \in L_0$ for all $t \ge 0$. Using Lemma 5.8 and its notation, we have

$$Z_t \overset{d}{=} b^{-1}Z_t + Z_t^{(b)},$$

where $b^{-1}Z_t$ and $Z_t^{(b)}$ are independent. Then,

$$Ee^{i\langle z,Y_t\rangle} = Ee^{i\langle b^{-1/\alpha}z,Y_t\rangle}E\left[\widehat{\mu}_{Z_t^{(b)}}(z)\right]. \tag{5.2}$$

Indeed we have, using Lemma 4.21 (i) and Lemma 5.7,

$$\begin{aligned} Ee^{i\langle z,Y_t\rangle} &= E\left[\left(Ee^{i\langle z,X_s\rangle}\right)_{s=Z_t}\right] = E\left[\widehat{\mu}_{Z_t}(z)\right] = E\left[\widehat{\mu}_{b^{-1}Z_t+Z_t^{(b)}}(z)\right] \\ &= E\left[\widehat{\mu}_{b^{-1}Z_t}(z)\widehat{\mu}_{Z_t^{(b)}}(z)\right] = E\left[\widehat{\mu}_{b^{-1}Z_t}(z)\right]E\left[\widehat{\mu}_{Z_t^{(b)}}(z)\right] \\ &= E\left[\widehat{\mu}_{Z_t}\left(b^{-1/\alpha}z\right)\right]E\left[\widehat{\mu}_{Z_t^{(b)}}(z)\right], \end{aligned}$$

which is the right-hand side of (5.2). Notice that $b^{1/\alpha}$ can be an arbitrary real bigger than 1 and $E\left[\widehat{\mu}_{Z_t^{(b)}}(z)\right]$ is the characteristic function of a subordinated process by Lemma 5.8. This shows that $\{Y_t\}$ is selfdecomposable.

(ii) By induction. If $m = 0$, then the assertion is true by (i). Suppose that the assertion is true for $m-1$ in place of m. Let $\{Z_t\}$ be of class L_m, that is, $\mathcal{L}(Z_t) \in L_m$ for $t \ge 0$. Then $\left\{Z_t^{(b)}\right\}$ is a K-valued subordinator of class L_{m-1} by Lemma 5.8. Hence $E\left[\widehat{\mu}_{Z_t^{(b)}}(z)\right]$ is a characteristic function of class L_{m-1}. Thus $\mathcal{L}(Y_t) \in L_m$.

(iii) Let $\mathcal{L}(Z_t) \in \mathfrak{S}^0_\beta$ for $t \ge 0$. Then $Z_{at} \overset{d}{=} a^{1/\beta}Z_t$. Therefore, using Lemma 5.7,

$$\begin{aligned} Ee^{i\langle z,Y_{at}\rangle} &= E\left[\left(Ee^{i\langle z,X_s\rangle}\right)_{s=Z_{at}}\right] = E\left[\left(Ee^{i\langle z,X_s\rangle}\right)_{s=a^{1/\beta}Z_t}\right] \\ &= E\left[\widehat{\mu}_{a^{1/\beta}Z_t}(z)\right] = E\left[\widehat{\mu}_{Z_t}\left(a^{1/(\alpha\beta)}z\right)\right] = E\left[e^{i\langle z,a^{1/(\alpha\beta)}Y_t\rangle}\right]. \end{aligned}$$

Thus $Y_{at} \overset{d}{=} a^{1/(\alpha\beta)}Y_t$ for any $a > 0$. ■

When $d = 1$, Theorem 5.1 can be generalized to the case where $\{X_t : t \geq 0\}$ is Brownian motion with non-zero drift on $\mathbb{R}$. This is 2-stable, but not strictly 2-stable. So the assumption in Theorem 5.1 is not satisfied. Nevertheless, selfdecomposability is inherited as follows.

Theorem 5.10 *Let $\{X_t : t \geq 0\}$ be Brownian motion with drift γ on $\mathbb{R}$. That is,*

$$Ee^{izX_t} = e^{t(-(z^2/2)+i\gamma z)}, \qquad z \in \mathbb{R}.$$

Let $\{Y_t\}$ be a Lévy process subordinate to $\{X_t\}$ by $\{Z_t\}$. If $\{Z_t\}$ is selfdecomposable, then $\{Y_t\}$ is selfdecomposable.

See Sato [94] (2001a).

Remark 5.11 There arises the question whether Theorem 5.10 can be extended to the case where $\{X_t\}$ is an α-stable, not strictly α-stable process with $0 < \alpha < 2$ on $\mathbb{R}$. Ramachandran's paper [80] (1997) contains an answer to this question.[1] Namely, if $1 < \alpha < 2$, then there are an α-stable, not strictly α-stable process $\{X_t\}$ on $\mathbb{R}$ and a selfdecomposable subordinator $\{Z_t\}$ such that the Lévy process $\{Y_t\}$ subordinate to $\{X_t\}$ by $\{Z_t\}$ is not selfdecomposable. Specifically, Ramachandran shows that if $Ee^{izX_t} = e^{t(-c|z|^\alpha + i\gamma z)}$ with $1 < \alpha < 2$, $c > 0$, and $\gamma \neq 0$ and $\{Z_t\}$ is Γ-process with parameter $q > 0$ (a special case of Example 4.7), then $\{Y_t\}$ is not selfdecomposable. The question in the case $0 < \alpha \leq 1$ is still open in the authors' knowledge.

Remark 5.12 If $d \geq 2$, then the situation is quite different and Theorem 5.10 cannot be generalized. It is known that, for $d \geq 2$, a Lévy process $\{Y_t\}$ on $\mathbb{R}^d$ subordinate to Brownian motion with drift, $\{X_t\}$, by a selfdecomposable subordinator $\{Z_t\}$ is not necessarily selfdecomposable. Even if $\mathcal{L}(Z_1)$ is a generalized Γ-convolution, $\{Y_t\}$ is not necessarily selfdecomposable.

Definition 5.13 The distribution

$$\mu(dx) = c \exp\left(-a\sqrt{1+x^2} + bx\right) dx$$

on $\mathbb{R}$ with parameters a, b satisfying $a > 0$ and $|b| < a$ or a scale change of this distribution is called *hyperbolic distribution*. Here c is a normalizing constant.

The distribution

$$\mu(dx) = c\left(\sqrt{1+x^2}\right)^{\lambda-(1/2)} K_{\lambda-(1/2)}\left(a\sqrt{1+x^2}\right) e^{bx}$$

on $\mathbb{R}$ or its scale change, where c is normalizing constant, is called *generalized hyperbolic distribution*. Here the domain of parameters is given by $\{\lambda \geq 0,\ a >$

[1] Z. J. Jurek kindly called the authors' attention to the paper [80] on this point.

$0,\ |b| < a\}$ and $\{\lambda < 0,\ a > 0,\ |b| \le a\}$. This distribution reduces to the hyperbolic distribution if $\lambda = 1$.

Example 5.14 Let $\{X_t\}$ be Brownian motion with drift γ being zero or non-zero and let $\{Z_t\}$ the subordinator with $\mathcal{L}(Z_1)$ being generalized inverse Gaussian $\mu_{\lambda,\chi,\psi}$ with $\lambda = 1$, $\chi > 0$, $\psi > 0$. Let us calculate the distribution at $t = 1$ for the Lévy process $\{Y_t\}$ subordinate to $\{X_t\}$ by $\{Z_t\}$:

$$P[Y_1 \in B] = c\int_0^\infty e^{-(\chi s^{-1}+\psi s)/2} ds \int_B \frac{1}{\sqrt{2\pi s}} e^{-(x-s\gamma)^2/(2s)} dx$$
$$= \frac{c}{\sqrt{\psi+\gamma}} \int_B e^{-\sqrt{(\psi+\gamma)(\chi+x^2)}+\gamma x} dx$$

by the calculation in Example 2.13 of [93]. Hence $\mathcal{L}(Y_1)$ is a hyperbolic distribution with $a = \sqrt{\chi(\psi+\gamma)}$ and $b = \sqrt{\chi}\gamma$.

More generally if we assume that $\mathcal{L}(Z_1)$ is generalized inverse Gaussian $\mu_{\lambda,\chi,\psi}$, then $\mathcal{L}(Y_1)$ is generalized hyperbolic distribution. For a proof, use the formula (30.28) of [93] for modified Bessel functions. It follows from Theorem 5.1 (if $\gamma = 0$) and Theorem 5.10 (if $\gamma \neq 0$) that generalized hyperbolic distributions are selfdecomposable.

5.2 Operator Generalization

For distributions on $\mathbb{R}^d$, $d \ge 2$, the concepts of stability, selfdecomposability, and L_m property are generalized to the situation where multiplication by positive real numbers is replaced by multiplication by matrices of the form b^Q.

For a set $J \subset \mathbb{R}$ let $M_J(d)$ be the set of real $d \times d$ matrices all of whose eigenvalues have real parts in J. Let $Q \in M_{(0,\infty)}(d)$.

Definition 5.15 A distribution μ on $\mathbb{R}^d$ is called *Q-selfdecomposable* if, for every $b > 1$, there is $\rho_b \in \mathfrak{P}(\mathbb{R}^d)$ such that

$$\widehat{\mu}(z) = \widehat{\mu}(b^{-Q^\top} z)\widehat{\rho}_b(z), \qquad z \in \mathbb{R}^d, \tag{5.3}$$

where $Q^\top$ is the transpose of Q and $b^{-Q^\top}$ is a $d \times d$ matrix defined by

$$b^{-Q^\top} = e^{-(\log b)Q^\top} = \sum_{n=0}^\infty (n!)^{-1}(-\log b)^n (Q^\top)^n.$$

The class of all Q-selfdecomposable distributions on $\mathbb{R}^d$ is denoted by $L_0(Q)$. For $m = 1, 2, \ldots$ the class $L_m(Q)$ is defined to be the class of distributions μ on $\mathbb{R}^d$

such that, for every $b > 1$, there exists $\rho_b \in L_{m-1}(Q)$ satisfying (5.3). Define $L_\infty(Q) = \bigcap_{m<\infty} L_m(Q)$.

It follows that $L_m(Q) = L_m(aQ)$ for any $a > 0$ and $m = 0, 1, \ldots, \infty$.

Proposition 5.16 *The classes just introduced form nested classes*

$$ID \supset L_0(Q) \supset L_1(Q) \supset \cdots \supset L_\infty(Q). \tag{5.4}$$

Proof can be given analogously to the proofs of Propositions 1.13 and 1.15. See Jurek [39] (1983a) and Sato and Yamazato [111] (1985).

Definition 5.17 A distribution μ on $\mathbb{R}^d$ is called *Q-stable* if, for every $n \in \mathbb{N}$, there is $c \in \mathbb{R}^d$ such that

$$\widehat{\mu}(z)^n = \widehat{\mu}(n^{Q^\top} z) e^{i\langle c,z\rangle}, \qquad z \in \mathbb{R}^d. \tag{5.5}$$

It is called *strictly Q-stable* if, for all n,

$$\widehat{\mu}(z)^n = \widehat{\mu}(n^{Q^\top} z), \qquad z \in \mathbb{R}^d. \tag{5.6}$$

Let $\mathfrak{S}_Q$ be the class of Q-stable distributions on $\mathbb{R}^d$. Let $\mathfrak{S}_Q^0$ be the class of strictly Q-stable distributions on $\mathbb{R}^d$.

Here we are using the usual terminology, but it is not harmonious with the usage of the word α-stable; μ is α-stable if and only if it is $(\alpha^{-1}I)$-stable, where I is the identity matrix. Similarly to the α-stable case, we have the following.

Proposition 5.18 *A distribution μ is Q-stable if and only if $\mu \in ID$ and, for every $t > 0$, there is $c \in \mathbb{R}^d$ such that*

$$\widehat{\mu}(z)^t = \widehat{\mu}(t^{Q^\top} z) e^{i\langle c,z\rangle}. \tag{5.7}$$

A distribution μ is strictly Q-stable if and only if $\mu \in ID$ and, for every $t > 0$,

$$\widehat{\mu}(z)^t = \widehat{\mu}(t^{Q^\top} z). \tag{5.8}$$

Proof is like that of Proposition 1.21.

Remark 5.19 If $\mu \in \mathfrak{S}_Q$ for some $Q \in M_{(0,\infty)}(d)$, then μ is called *operator stable* and sometimes Q is called exponent of operator stability of μ. But, in general, Q is not uniquely determined by μ; see Hudson and Mason [35] (1981) and Sato [86] (1985). If $\mu \in L_0(Q)$ for some $Q \in M_{(0,\infty)}(d)$, then μ is called *operator selfdecomposable*.

Remark 5.20 Operator stable and operator selfdecomposable distributions appear in a natural way when we study limit theorems for sums of a sequence of independent random vectors, allowing linear transformations (matrix multiplications) of partial sums. Basic papers are Sharpe [113] (1969) and Urbanik [128] (1972a).

Proposition 5.21 *Suppose that μ is Q-stable and nondegenerate on $\mathbb{R}^d$. Then Q must be in $M_{[1/2,\infty)}(d)$ and, moreover, any eigenvalue of Q with real part $1/2$ is a simple root of the minimal polynomial of Q; μ is Gaussian if and only if $Q \in M_{\{1/2\}}(d)$; μ is purely non-Gaussian if and only if $Q \in M_{(1/2,\infty)}(d)$.*

This is by Sharpe [113] (1969).

Definition 5.22 For $Q \in M_{(0,\infty)}(d)$, let $\mathfrak{S}(Q)$ denote the union of $\mathfrak{S}_{aQ}$ over all $a > 0$; let $\mathfrak{S}^0(Q)$ denote the union of $\mathfrak{S}^0_{aQ}$ over all $a > 0$. The relation with $\mathfrak{S}$ and $\mathfrak{S}^0$ in Definition 1.19 is that $\mathfrak{S} = \mathfrak{S}(I)$ and $\mathfrak{S}^0 = \mathfrak{S}^0(I)$.

The class $\mathfrak{S}(Q)$ is a subclass of $L_\infty(Q)$. Moreover, we have the following.

Proposition 5.23 *The class $L_\infty(Q)$ is the smallest class containing $\mathfrak{S}(Q)$ and closed under convolution and weak convergence.*

See Sato and Yamazato [111] (1985) for a proof.

Definition 5.24 A Lévy process $\{X_t : t \geq 0\}$ is called *Q-selfdecomposable*, *Q-stable*, or of *class $L_m(Q)$*, respectively, if $\mathcal{L}(X_1)$ (or, equivalently, $\mathcal{L}(X_t)$ for every $t \geq 0$) is Q-selfdecomposable, Q-stable, or of class $L_m(Q)$.

Here are results on the inheritance of operator selfdecomposability, $L_m(Q)$ property, and strict operator stability in some cases. These partially extend Theorem 5.9. Propositions 5.21 and 5.23 are not used in the proof.

Let N and d be positive integers satisfying $d \geq N \geq 1$. Let d_j, $1 \leq j \leq N$, be positive integers such that $d_1 + \cdots + d_N = d$. Every $x \in \mathbb{R}^d$ is expressed as $x = (x_j)_{1 \leq j \leq N}$ with $x_j \in \mathbb{R}^{d_j}$. We call x_j the jth *component-block* of x. The jth component-block of X_t is denoted by $(X_t)_j$. As in Sect. 4.3, we use the unit vectors $e^k = (\delta_{kj})_{1 \leq j \leq N}$, $k = 1, \ldots, N$, in $\mathbb{R}^N$.

Theorem 5.25 *Suppose that $\{X_s : s \in \mathbb{R}^N_+\}$ is a given $\mathbb{R}^N_+$-parameter Lévy process on $\mathbb{R}^d$ with the following structure: for each $j = 1, \ldots, N$,*

$$(X_{te^j})_k = 0 \quad \text{for all } k \neq j. \tag{5.9}$$

Suppose that $\{Z_t : t \geq 0\}$ is a given $\mathbb{R}^N_+$-valued subordinator and let $\{Y_t : t \geq 0\}$ be a Lévy process on $\mathbb{R}^d$ obtained by multivariate subordination from $\{X_s\}$ and $\{Z_t\}$. That is, $\{X_s\}$ and $\{Z_t\}$ are independent and $Y_t = X_{Z_t}$. Let $Q_j \in M_{[1/2,\infty)}(d_j)$ and $c_j > 0$ for $1 \leq j \leq N$, and let $C = \operatorname{diag}(c_1, \ldots, c_N)$. Assume that, for each j, $\mathcal{L}((X_{te^j})_j)$ is strictly Q_j-stable. Let

$$D = \operatorname{diag}(c_1 Q_1, \ldots c_N Q_N) \in M_{(0,\infty)}(d).$$

(i) If $\{Z_t : t \geq 0\}$ is C-selfdecomposable, then $\{Y_t : t \geq 0\}$ is D-selfdecomposable.

(ii) More generally, let $m \in \{0, 1, \ldots, \infty\}$. If $\{Z_t : t \geq 0\}$ is of class $L_m(C)$ on $\mathbb{R}^N$, then $\{Y_t : t \geq 0\}$ is of class $L_m(D)$ on $\mathbb{R}^d$.

(iii) If $\{Z_t : t \geq 0\}$ is strictly C-stable, then $\{Y_t : t \geq 0\}$ is strictly D-stable.

Here $\operatorname{diag}(c_1, \ldots, c_N)$ denotes the diagonal matrix with diagonal entries $c_1, \ldots, c_N$; $\operatorname{diag}(c_1 Q_1, \ldots c_N Q_N)$ denotes the blockwise diagonal matrix with diagonal blocks $c_1 Q_1, \ldots, c_N Q_N$.

Proof We use Theorem 4.41. Let $X_t^j = X_{te^j}$. Let $\psi_X^j(z) = (\log \widehat{\rho})(z)$ with $\rho = \mathcal{L}(X_1^j)$ for $z \in \mathbb{R}^d$, and $\psi_X(z) = (\psi_X^j(z))_{1 \leq j \leq N}$. Let $\mu_j = \mathcal{L}((X_1^j)_j) \in \mathfrak{P}(\mathbb{R}^{d_j})$. Then it follows from (5.9) that

$$e^{t\psi_X^j(z)} = Ee^{i\langle z, X_t^j \rangle} = Ee^{i\langle z_j, (X_t^j)_j \rangle} = \widehat{\mu}_j(z_j)^t,$$

where $z = (z_j)_{1 \leq j \leq N} \in \mathbb{R}^d$ with $z_j \in \mathbb{R}^{d_j}$. Thus

$$\psi_X(z) = (\log \widehat{\mu}_j(z_j))_{1 \leq j \leq N}.$$

We have

$$\widehat{\mu}_j(z_j)^a = \widehat{\mu}_j(a^{Q_j^\top} z_j), \qquad a > 0$$

by the strict Q_j-stability of μ_j. Hence

$$a^C \psi_X(z) = (a^{c_j} \log \widehat{\mu}_j(z_j))_{1 \leq j \leq N} = (\log \widehat{\mu}_j(a^{c_j Q_j^\top} z_j))_{1 \leq j \leq N}. \tag{5.10}$$

(i) Assume $\{Z_t : t \geq 0\}$ is C-selfdecomposable. Let Ψ_Z be the function Ψ in (4.11) for $\{Z_t\}$. For $b > 1$ and $w = (w_j)_{1 \leq j \leq N} \in \mathbb{C}^N$ with $\operatorname{Re} w_j \leq 0$, Define $\Psi_{Z,b}(w)$ by

$$\Psi_Z(w) = \Psi_Z(b^{-C} w) + \Psi_{Z,b}(w).$$

Similarly to the proof of Proposition 1.13, we can show that $e^{\Psi_{Z,b}(iu)}$, $u \in \mathbb{R}^N$, is an infinitely divisible characteristic function. Further, as in Lemma 5.8, there is an $\mathbb{R}_+^N$-valued subordinator $\{Z_t^{(b)}\}$ such that $Ee^{i\langle u, Z_t^{(b)} \rangle} = e^{t\Psi_{Z,b}(iu)}$. In the proof note that $\gamma_b^0 = (I - b^{-C})\gamma^0 = \operatorname{diag}(1 - b^{-c_1}, \ldots, 1 - b^{-c_N})\gamma^0 \in \mathbb{R}_+^N$. Now we have

$$Ee^{i\langle z, Y_t \rangle} = e^{t\Psi_Z(\psi_X(z))} = e^{t\Psi_Z(b^{-C}\psi_X(z))} e^{t\Psi_{Z,b}(\psi_X(z))}$$

and

$$b^{-C}\psi_X(z) = (\log \widehat{\mu}_j(b^{-c_j Q_j^\top} z_j))_{1 \leq j \leq N} = \psi_X(b^{-D^\top} z)$$

by (5.10), since

$$b^{-D^\top} z = \mathrm{diag}(b^{-c_1 Q_1^\top}, \ldots, b^{-c_N Q_N^\top}) z = (b^{-c_j Q_j^\top} z_j)_{1 \le j \le N}.$$

Hence

$$Ee^{i\langle z, Y_t\rangle} = \left(E \exp(i\langle b^{-D^\top} z, Y_t\rangle)\right) e^{t\Psi_{Z,b}(\psi_X(z))}.$$

As the second factor in the right-hand side is the characteristic function of a subordinated process, we see that $\mathcal{L}(Y_t)$ is D-selfdecomposable.

(ii) By induction similar to (ii) of Theorem 5.9.

(iii) Assume that $\{Z_t\}$ is strictly C-stable, that is, $a\Psi_Z(w) = \Psi_Z(a^C w)$. Then, for $a > 0$,

$$Ee^{i\langle z, Y_{at}\rangle} = e^{at\Psi_Z(\psi_X(z))} = e^{t\Psi_Z(a^C \psi_X(z))}$$

and, as above,

$$a^C \psi_X(z) = \psi_X(a^{D^\top} z).$$

Hence

$$Ee^{i\langle z, Y_{at}\rangle} = E \exp(i\langle a^{D^\top} z, Y_t\rangle),$$

which shows D-stability of $\{Y_t\}$. ■

Remark 5.26 Let $Q \in M_{(0,\infty)}(d)$ and let

$$S_Q = \{\xi \in \mathbb{R}^d : |\xi| = 1, \text{ and } |r^Q \xi| > 1 \text{ for every } r > 1\}.$$

Then any $x \in \mathbb{R}^d \setminus \{0\}$ is uniquely expressed as $x = r^Q \xi$ with $r > 0$ and $\xi \in S_Q$. Notice that S_I is the unit sphere S but $S_Q \subsetneqq S$ for some Q. Let $\mu \in ID$ with generating triplet (A, ν, γ). Then $\mu \in L_0(Q)$ if and only if $QA + AQ^\top$ is nonnegative-definite and

$$\nu(B) = \int_{S_Q} \lambda(d\xi) \int_0^\infty 1_B(r^Q \xi) \frac{k_\xi(r)}{r} dr, \qquad B \in \mathcal{B}(\mathbb{R}^d),$$

where λ is a finite measure on S_Q and $k_\xi(r)$ is nonnegative, decreasing in $r \in (0, \infty)$, and measurable in $\xi \in S_Q$. Under the assumption that $\alpha > 0$, $Q \in M_{(\alpha/2,\infty)}(d)$, and μ is purely non-Gaussian, we can show that $\mu \in \mathfrak{S}_{\alpha^{-1}Q}$ if and only if

$$\nu(B) = \int_{S_Q} \lambda(d\xi) \int_0^\infty 1_B(r^Q \xi) r^{-\alpha-1} dr, \qquad B \in \mathcal{B}(\mathbb{R}^d),$$

where λ is a finite measure on S_Q; this statement does not exclude the possibility that $\mathfrak{S}_{\alpha^{-1}Q}$ is the set of trivial distributions. It follows that, if $\{Z_t\}$ is a non-trivial $(\alpha^{-1}Q)$-stable $\mathbb{R}_+^d$-valued subordinator, then Q is strongly restricted. For example then, under the additional assumption that $d = 2$ and Q is of the real Jordan normal form, Q cannot be

$$\begin{pmatrix} q_1 & 1 \\ 0 & q_1 \end{pmatrix} \quad \text{nor} \quad \begin{pmatrix} q_1 & -q_2 \\ q_2 & q_1 \end{pmatrix}$$

with $q_1 > 0$, $q_2 > 0$ and thus Q must be of the form

$$\begin{pmatrix} q_1 & 0 \\ 0 & q_2 \end{pmatrix}$$

with $q_1 > 0$, $q_2 > 0$. See Sato [86, pp. 42–43] (1985), Sato and Yamazato [111] (1985), and Jurek and Mason [44] (1993).

Notes

Halgreen [30] (1979) and Ismail and Kelker [36] (1979) proved assertion (i) of Theorem 5.1 in the case where $\{X_t\}$ is Brownian motion on $\mathbb{R}$. Assertion (iii) of Theorem 5.1 was essentially known to Bochner [14] (1955). Theorem 5.25 was given in Barndorff-Nielsen et al. [8] (2001), but we have given a simpler proof. Assertion (ii) of Theorem 5.1 is a special case of Theorem 5.25 (ii) with $N = 1$ and $Q = Q_1 = (1/\alpha)I$. Theorem 5.9 was shown by Pedersen and Sato [71] (2003) for subordination of cone-parameter convolution semigroups on $\mathbb{R}^d$.

Examples 5.5 and 5.14 are from Barndorff-Nielsen and Halgreen [6] (1977) and Halgreen [30] (1979); see also Bondesson [16] (1992).

Theorems 5.9 and 5.10 were extended to subordination of cone-parameter convolution semigroups, respectively, by Pedersen and Sato [71] (2003) and by Sato [100] (2009).

Theorem 5.10 was proved in Sato [94] (2001a). Earlier Halgreen [30] (1979) and Shanbhag and Sreehari [112] (1979) proved it under the condition that $\mathcal{L}(Z_1)$ is a generalized Γ-convolution. Remark 5.12 is by Takano [119] (1989,1990).

Characterization and many related results on general distributions in $\mathfrak{S}(Q)$ and $L_m(Q)$ are discussed in Sharpe [113] (1969), Urbanik [128] (1972a), Sato and Yamazato [111] (1985), and Jurek and Mason [44] (1993). For characterization of distributions in $\mathfrak{S}^0(Q)$, see Sato [87] (1987).

For $Q \in M_{(0,\infty)}(d)$ with $d \geq 2$, consider Eq. (2.13) with c replaced by Q. Then we can extend the notion of Ornstein–Uhlenbeck type process generated by $\rho \in ID(\mathbb{R}^d)$ and $c > 0$ to that generated by ρ and Q in a natural way; the extended process is also called Ornstein–Uhlenbeck type process frequently, but let us call it as Q-OU type process. Connections of distributions in $L_m(Q)$, $m = 0, 1, \ldots, \infty$, with Q-OU type processes are parallel to those of L_m with OU type processes in Chaps. 2 and 3 and the proofs are similar; in fact it was done simultaneously in many papers. However, it was a harder problem to find a criterion of recurrence and transience for Q-OU type processes; it was solved by Sato et al. [104] (1996) and Watanabe [133] (1998).

Bibliography

1. Akita, K., & Maejima, M. (2002). On certain self-decomposable self-similar processes with independent increments. *Statistics & Probability Letters, 59*, 53–59.
2. Alf, C., & O'Connor, T. A. (1977). Unimodality of the Lévy spectral function. *Pacific Journal of Mathematics, 69*, 285–290.
3. Araujo, A., & Giné, E. (1980). *The central limit theorem for real and banach valued random variables*. New York, Wiley.
4. Barndorff-Nielsen, O. E. (1978). Hyperbolic distributions and distributions on hyperbolae. *Scandinavian Journal of Statistics, 5*, 151–157.
5. Barndorff-Nielsen, O. E. (1997). Normal inverse Gaussian distributions and stochastic volatility modelling. *Scandinavian Journal of Statistics, 24*, 1–13.
6. Barndorff-Nielsen, O. E. & Halgreen, C. (1977). Infinite divisibility of the hyperbolic and generalized inverse Gaussian distributions. *Zeitschrift für Wahrscheinlichkeitstheorie und verwandte Gebiete, 38*, 309–311.
7. Barndorff-Nielsen, O. E., Maejima, M., & Sato, K. (2006). Some classes of infinitely divisible distributions admitting stochastic integral representations. *Bernoulli, 12*, 1–33.
8. Barndorff-Nielsen, O. E., Pedersen, J., & Sato, K. (2001). Multivariate subordination, selfdecomposability and stability. *Advances in Applied Probability, 33*, 160–187.
9. Barndorff-Nielsen, O. E., & Pérez-Abreu, V. (2008). Matrix subordinators and related upsilon transformations. *Theory of Probability and its Applications, 52*, 1–23.
10. Barndorff-Nielsen, O. E., Rosiński, J., & Thorbjørnsen, S. (2008). General Υ transformations. *ALEA Latin American Journal of Probability and Mathematical Statistics, 4*, 131–165.
11. Barndorff-Nielsen, O. E., & Thorbjørnsen, S. (2004). A connection between free and classical infinite divisibility. *Infinite Dimensional Analysis, Quantum Probability and Related Topics, 7*, 573–590.
12. Bertoin, J. (1996). *Lévy processes*. Cambridge: Cambridge University Press.
13. Billingsley, P. (2012). *Probability and measure* (Anniversary ed.). New Jersey: Wiley.
14. Bochner, S. (1955). *Harmonic analysis and the theory of probability*. Berkeley: University of California Press.
15. Bondesson, L. (1981). Classes of infinitely divisible distributions and densities. *Zeitschrift für Wahrscheinlichkeitstheorie und verwandte Gebiete, 57*, 39–71. Correction and addendum 59, 277 (1982).
16. Bondesson, L. (1992). *Generalized gamma convolutions and related classes of distribution densities. Lecture notes in statistics* (Vol. 76). New York: Springer.

A. Rocha-Arteaga, K. Sato, *Topics in Infinitely Divisible Distributions and Lévy Processes*, SpringerBriefs in Probability and Mathematical Statistics,
https://doi.org/10.1007/978-3-030-22700-5

17. Carr, P., Geman, H., Madan, D. B, & Yor, M. (2005). Pricing options on realized variance. *Finance and Stochastics, 9*, 453–475.
18. Chentsov, N. N. (1957). Lévy's Brownian motion of several parameters and generalized white noise. *Theory of Probability and its Applications, 2*, 265–266.
19. Chung, K. L. (1974). *A course in probability theory* (2nd ed.). Orlando, FL: Academic Press.
20. Çinlar, E., & Pinsky, M. (1971). A stochastic integral in storage theory. *Zeitschrift für Wahrscheinlichkeitstheorie und verwandte Gebiete, 17*, 227–240.
21. Çinlar, E., & Pinsky, M. (1972). On dams with additive inputs and a general release rule. *Journal of Applied Probability, 9*, 422–429.
22. Dettweiler, E. (1976). Infinitely divisible measures on the cone of an ordered locally convex vector space. *Annales scientifiques de l'Université de Clermont, 14*, 61, 11–17.
23. Eberlein, E., & Madan, D. B. (2009). Sato processes and the valuation of structured products. *Quantitative Finance, 9*, 1, 27–42.
24. Feller, W. (1971). *An introduction to probability theory and its applications* (2nd ed., Vol. 2). New York: Wiley.
25. Gihman, I. I., & Skorohod, A. V. (1975). *The theory of stochastic processes* (Vol. II). Berlin: Springer.
26. Gnedenko, B. V., & Kolmogorov, A. N. (1968). *Limit distributions for sums of independent random variables* (2nd ed.). Reading, MA: Addison-Wesley (Russian original 1949).
27. Goldie, C. (1967). A class of infinitely divisible random variables. *Proceedings of the Cambridge Philosophical Society, 63*, 1141–1143.
28. Gravereaux, J. B. (1982). Probabilité de Lévy sur $\mathbb{R}^d$ et équations différentielles stochastiques linéaires, *Séminaire de Probabilité 1982* (Université de Rennes I, Publications des Séminaires de Mathématiques), 1–42.
29. Graversen, S. E., & Pedersen, J. (2011). Representations of Urbanik's classes and multiparameter Ornstein–Uhlenbeck processes. *Electronic Communications in Probability, 16*, 200–212.
30. Halgreen, C. (1979). Self-decomposability of the generalized inverse Gaussian and hyperbolic distributions. *Zeitschrift für Wahrscheinlichkeitstheorie und verwandte Gebiete, 47*, 13–17.
31. Halmos, P. R. (1950). *Measure theory*. Princeton, NJ: Van Nostrand.
32. Hartman, P., & Wintner, A. (1942). On the infinitesimal generators of integral convolutions. *American Journal of Mathematics, 64*, 273–298.
33. Hengartner, W., & Theodorescu, R. (1973). *Concentration functions*. New York: Academic Press.
34. Hirsch, F. (1995). Potential theory related to some multiparameter processes. *Potential Analysis, 4*, 245–267.
35. Hudson, W. N., & Mason, J. D. (1981). Operator-stable distribution on R^2 with multiple exponents. *Annals of Probability, 9*, 482–489.
36. Ismail, M. E. H., & Kelker, D. H. (1979). Special functions, Stieltjes transforms and infinite divisibility. *SIAM Journal on Mathematical Analysis, 10*, 884–901.
37. James, L. F., Roynette, B., & Yor, M. (2008). Generalized gamma convolutions, Dirichlet means, Thorin measures, with explicit examples. *Probability Surveys, 5*, 346–415.
38. Jeanblanc, M., Pitman, J., & Yor, M. (2002). Self-similar processes with independent increments associated with Lévy and Bessel processes. *Stochastic Processes and their Applications, 100*, 223–231.
39. Jurek, Z. J. (1983a). Limit distributions and one-parameter groups of linear operators on Banach spaces. *Journal of Multivariate Analysis, 13*, 578–604.
40. Jurek, Z. J. (1983b). The class $L_m(Q)$ of probability measures on Banach spaces. *Bulletin of the Polish Academy of Sciences Mathematics, 31*, 51–62.
41. Jurek, Z. J. (1985). Relations between the s-selfdecomposable and selfdecomposable measures. *Annals of Probability, 13*, 592–608.
42. Jurek, Z. J. (1988). Random integral representations for classes of limit distributions similar to Levy class L_0. *Probability Theory and Related Fields, 78*, 473–490.
43. Jurek, Z. J. (1989). Random integral representations for classes of limit distributions similar to Lévy class L_0, II. *Nagoya Mathematical Journal, 114*, 53–64.

44. Jurek, Z. J., & Mason, J. D. (1993). *Operator-limit distributions in probability theory*. New York: Wiley.
45. Jurek, Z. J., & Vervaat, W. (1983). An integral representation for self-decomposable Banach space valued random variables. *Zeitschrift für Wahrscheinlichkeitstheorie und verwandte Gebiete, 62*, 247–262.
46. Khintchine, A. Ya. (1938). *Limit laws for sums of independent random variables*. Moscow: ONTI (in Russian).
47. Khoshnevisan, D., & Shi, Z. (1999). Brownian sheet and capacity. *Annals of Probability, 27*, 1135–1159.
48. Kokholm, T., & Nicolato, E. (2010). Sato processes in default modelling. *Applied Mathematical Finance, 17*(5), 377–397.
49. Kumar, A., & Schreiber, B. M. (1978). Characterization of subclasses of class L probability distributions. *Annals of Probability, 6*, 279–293.
50. Kumar, A., & Schreiber, B. M. (1979). Representation of certain infinitely divisible probability measures on Banach spaces. *Journal of Multivariate Analysis, 9*, 288–303.
51. Kyprianou, A. E. (2014). *Fluctuations of Lévy processes with applications* (2nd ed.). Berlin: Springer.
52. Lagaize, S. (2001). Hölder exponent for a two-parameter Lévy process. *Journal of Multivariate Analysis, 77*, 270–285.
53. Lamperti, J. (1962). Semi-stable stochastic processes. *Transactions of the American Mathematical Society, 104*, 62–78.
54. Lévy, P. (1937). *Théorie de l'Addition des Variables Aléatoires*. Paris: Gauthier-Villars. (2 éd. 1954).
55. Lévy, P. (1948). *Processus Stochastiques et Mouvement Brownien*. Paris: Gauthier-Villars. (2nd éd. 1965).
56. Linde, W. (1986). *Probability in Banach spaces -stable and infinitely divisible distributions*. Berlin: Wiley.
57. Linnik, J. V., & Ostrovskii, I. V. (1977). *Decomposition of random variables and vectors*. Providence, RI: American Mathematical Society.
58. Loève, M. (1977, 1978). *Probability theory* (4th ed., Vols. I and II). New York: Springer (1st ed., 1955).
59. Maejima, M. (2015). Classes of infinitely divisible distributions and examples. In *Lévy Matters V. Lecture notes in mathematics* (Vol. 2149, pp. 1–65). Cham: Springer
60. Maejima, M., & Sato, K. (2003). Semi-Lévy processes, semi-selfsimilar additive processes, and semi-stationary Ornstein–Uhlenbeck type processes. *Journal of Mathematics of Kyoto University, 43*, 609–639.
61. Maejima, M., & Sato, K. (2009). The limits of nested subclasses of several classes of infinitely divisible distributions are identical with the closure of the class of stable distributions. *Probability Theory and Related Fields, 145*, 119–142.
62. Maejima, M., Sato, K., & Watanabe, T. (2000). Distributions of selfsimilar and semi-selfsimilar processes with independent increments. *Statistics & Probability Letters, 47*, 395–401.
63. Maejima, M., Suzuki, K., & Tamura, Y. (1999). Some multivariate infinitely divisible distributions and their projections. *Probability and Mathematical Statistics, 19*, 421–428.
64. Maejima, M., & Ueda, Y. (2009). Stochastic integral characterizations of semi-selfdecomposable distributions and related Ornstein–Uhlenbeck type processes. *Communications on Stochastic Analysis, 3*, 349–367.
65. Maejima, M., & Ueda, Y. (2010). α-self-decomposable distributions and related Ornstein–Uhlenbeck type processes. *Stochastic Processes and their Applications, 120*, 2363–2389.
66. McKean, H. P., Jr. (1963). Brownian motion with a several dimensional time. *Theory of Probability and its Applications, 8*, 357–378.
67. O'Connor, T. A. (1979a). Infinitely divisible distributions with unimodal Lévy spectral functions. *Annals of Probability, 7*, 494–499.

68. O'Connor, T. A. (1979b) Infinitely divisible distributions similar to class L distributions. *Zeitschrift für Wahrscheinlichkeitstheorie und verwandte Gebiete, 50*, 265–271.
69. Orey, S. (1968). On continuity properties of infinitely divisible distribution functions. *Annals of Mathematical Statistics, 39*, 936–937.
70. Orey, S., & Pruitt, W. E. (1973). Sample functions of the N-parameter Wiener process. *Annals of Probability, 1*, 138–163.
71. Pedersen, J., & Sato, K. (2003). Cone-parameter convolution semigroups and their subordination. *Tokyo Journal of Mathematics, 26*, 503–525.
72. Pedersen, J., & Sato, K. (2004). Relations between cone-parameter Lévy processes and convolution semigroups. *Journal of the Mathematical Society of Japan, 56*, 541–559.
73. Pérez-Abreu, V., & Rocha-Arteaga, A. (2003). Lévy processes in Banach spaces: Distributional properties and subordination. *Stochastic Models, Contemporary Mathematics Series of the American Mathematical Society, 336*, 225–235.
74. Pérez-Abreu, V., & Rocha-Arteaga, A. (2005). Covariance-parameter Lévy processes in the space of trace-class operators. *Infinite Dimensional Analysis, Quantum Probability and Related Topics, 8*, 33–54.
75. Pérez-Abreu, V., & Rocha-Arteaga, A. (2006). On the Lévy–Khintchine representation of Lévy processes in cones of Banach spaces. *Publicaciones Matemáticas del Uruguay, 11*, 41–55.
76. Pérez-Abreu, V., Rocha-Arteaga, A., & Tudor, C. (2005). Cone-additive processes in duals of nuclear Fréchet spaces. *Random Operators and Stochastic Equations, 13*, 353–368.
77. Pérez-Abreu, V., & Rosiński, J. (2007). Representation of infinitely divisible distributions on cones. *Journal of Theoretical Probability, 20*, 535–544.
78. Pérez-Abreu, V., & Stelzer, R. (2014). Infinitely divisible multivariate and matrix gamma distributions. *Journal of Multivariate Analysis, 130*, 155–175.
79. Pillai, R. N. (1990). On Mittag–Leffler functions and related distributions. *Annals of the Institute of Statistical Mathematics, 42*, 157–161.
80. Ramachandran, B. (1997). On geometric-stable laws, a related property of stable processes, and stable densities of exponent one. *Annals of the Institute of Statistical Mathematics, 49*, 299–313.
81. Rocha-Arteaga, A. (2006). Subordinators in a class of Banach spaces. *Random Operators and Stochastic Equations, 14*, 1–14.
82. Rockafellar, B. T. (1970). *Convex analysis*. Princeton, NJ: Princeton University Press.
83. Rosiński, J. (2007). Tempering stable processes. *Stochastic Processes and their Applications, 117*, 677–707.
84. Sato, K. (1980). Class L of multivariate distributions and its subclasses. *Journal of Multivariate Analysis, 10*, 207–232.
85. Sato, K. (1982). Absolute continuity of multivariate distributions of class L. *Journal of Multivariate Analysis, 12*, 89–94.
86. Sato, K. (1985). *Lectures on multivariate infinitely divisible distributions and operator-stable processes*. Technical Report Series, Laboratory for Research in Statistics and Probability, Carleton University and University of Ottawa, No. 54, Ottawa.
87. Sato, K. (1987). Strictly operator-stable distributions. *Journal of Multivariate Analysis, 22*, 278–295.
88. Sato, K. (1990). Distributions of class L and self-similar processes with independent increments. In T. Hida, H. H. Kuo, J. Potthoff & L. Streit (Eds.), *White noise analysis. Mathematics and applications* (pp. 360–373). Singapore: World Scientific.
89. Sato, K. (1991). Self-similar processes with independent increments. *Probability Theory and Related Fields, 89*, 285–300.
90. Sato, K. (1994). Time evolution of distributions of Lévy processes from continuous singular to absolutely continuous. *Research Bulletin, College of General Education, Nagoya University, Series B, 38*, 1–11.

91. Sato, K. (1997). Time evolution of Lévy processes. In N. Kono & N.-R. Shieh (Eds.), *Trends in Probability and Related Analysis, Proceedings SAP '96* (pp. 35–82). Singapore: World Scientific.
92. Sato, K. (1998). Multivariate distributions with selfdecomposable projections. *Journal of the Korean Mathematical Society, 35*, 783–791.
93. Sato, K. (1999). *Lévy processes and infinitely divisible distributions*. Cambridge: Cambridge University Press.
94. Sato, K. (2001a). Subordination and selfdecomposability. *Statistics & Probability Letters, 54*, 317–324.
95. Sato, K. (2001b). Basic results on Lévy processes. In O. E. Barndorff-Nielsen, T. Mikosch & S. I. Resnick (Eds.), *Lévy processes, theory and applications* (pp. 3–37). Boston: Birkhäuser.
96. Sato, K. (2004). Stochastic integrals in additive processes and application to semi-Lévy processes. *Osaka Journal of Mathematics, 41*, 211–236.
97. Sato, K. (2006a). Additive processes and stochastic integrals. *Illinois Journal of Mathematics, 50*, 825–851.
98. Sato, K. (2006b). Two families of improper stochastic integrals with respect to Lévy processes. *ALEA Latin American Journal of Probability and Mathematical Statistics, 1*, 47–87.
99. Sato, K. (2006c). Monotonicity and non-monotonicity of domains of stochastic integral operators. *Probability and Mathematical Statistics, 26*, 23–39.
100. Sato, K. (2009). Selfdecomposability and semi-selfdecomposability in subordination of cone-parameter convolution semigroups. *Tokyo Journal of Mathematics, 32*, 81–90.
101. Sato, K. (2010). Fractional integrals and extensions of selfdecomposability. In *Lévy matters I. Lecture notes in mathematics* (Vol. 2001, pp. 1–91). Cham: Springer.
102. Sato, K. (2011). Description of limits of ranges of iterations of stochastic integral mappings of infinitely divisible distributions. *ALEA Latin American Journal of Probability and Mathematical Statistics, 8*, 1–17.
103. Sato, K. (2013). *Lévy processes and infinitely divisible distributions* (Revised ed.). Cambridge: Cambridge University Press.
104. Sato, K., Watanabe, T., Yamamuro, K., & Yamazato, M. (1996). Multidimensional process of Ornstein–Uhlenbeck type with nondiagonalizable matrix in linear drift terms. *Nagoya Mathematical Journal, 141*, 45–78.
105. Sato, K., Watanabe, T., & Yamazato, M. (1994). Recurrence conditions for multidimensional processes of Ornstein–Uhlenbeck type. *Journal of the Mathematical Society of Japan, 46*, 245–265.
106. Sato, K., & Yamamuro, K. (1998). On selfsimilar and semi-selfsimilar processes with independent increments. *Journal of the Korean Mathematical Society, 35*, 207–224.
107. Sato, K., & Yamamuro, K. (2000). Recurrence-transience for self-similar additive processes associated with stable distributions. *Acta Applicandae Mathematica, 63*, 375–384.
108. Sato, K., & Yamazato, M. (1978). On distribution functions of class *L*. *Zeitschrift für Wahrscheinlichkeitstheorie und verwandte Gebiete, 43*, 273–308.
109. Sato, K., & Yamazato, M. (1983). Stationary processes of Ornstein–Uhlenbeck type. In K. Itô & J. V. Prokhorov (Eds.), *Probability theory and mathematical statistics, Fourth USSR–Japan symposium, proceedings 1982. Lecture notes in mathematics* (No. 1021, pp. 541–551). Berlin: Springer.
110. Sato, K., & Yamazato, M. (1984). Operator-selfdecomposable distributions as limit distributions of processes of Ornstein–Uhlenbeck type. *Stochastic Processes and their Applications, 17*, 73–100.
111. Sato, K., & Yamazato, M. (1985). Completely operator-selfdecomposable distributions and operator-stable distributions. *Nagoya Mathematical Journal, 97*, 71–94.
112. Shanbhag, D. N., & Sreehari, M. (1979). An extension of Goldie's result and further results in infinite divisibility. *Zeitschrift für Wahrscheinlichkeitstheorie und verwandte Gebiete, 47*, 19–25.

113. Sharpe, M. (1969). Operator-stable probability distributions on vector groups. *Transactions of the American Mathematical Society, 136*, 51–65.
114. Shiga, T. (1990). A recurrence criterion for Markov processes of Ornstein–Uhlenbeck type. *Probability Theory and Related Fields, 85*, 425–447.
115. Skorohod, A. V. (1991). *Random processes with independent increments*. Dordrecht: Kluwer Academic Publications (Russian original 1986).
116. Steutel, F. W. (1967). Note on the infinite divisibility of exponential mixtures. *Annals of Mathematical Statistics, 38*, 1303–1305.
117. Steutel, F. W. (1970). *Preservation of infinite divisibility under mixing and related topics*, Math. Centre Tracs. No. 33, Math Centrum Amsterdam.
118. Steutel, F. W., & van Harn, K. (2004). *Infinite divisibility of probability distributions on the real line*. New York: Marcel Decker.
119. Takano, K. (1989/1990). On mixtures of the normal distribution by the generalized gamma convolutions. *Bulletin of the Faculty of Science, Ibaraki University. Series A, 21*, 29–41. Correction and addendum, **22**, 49–52.
120. Thorin, O. (1977a). On the infinite divisibility of the Pareto distribution. *Scandinavian Actuarial Journal, 1977*, 31–40.
121. Thorin, O. (1977b). On the infinite divisibility of the lognormal distribution. *Scandinavian Actuarial Journal, 1977*, 121–148 .
122. Thu, N. V. (1979). Multiply self-decomposable probability measures on Banach spaces. *Studia Mathematica, 66*, 161–175.
123. Thu, N. V. (1982). Universal multiply self-decomposable probability measures on Banach spaces. *Probability and Mathematical Statistics, 3*, 71–84.
124. Thu, N. V. (1984). Fractional calculus in probability. *Probability and Mathematical Statistics, 3*, 173–189.
125. Thu, N. V. (1986). An alternative approach to multiply self-decomposable probability measures on Banach spaces. *Probability Theory and Related Fields, 72*, 35–54.
126. Tucker, H. G. (1965). On a necessary and sufficient condition that an infinitely divisible distribution be absolutely continuous. *Transactions of the American Mathematical Society, 118*, 316–330.
127. Urbanik, K. (1969). Self-decomposable probability distributions on R^m. *Applicationes Mathematicae, 10*, 91–97.
128. Urbanik, K. (1972a). Lévy's probability measures on Euclidean spaces. *Studia Mathematica, 44*, 119–148.
129. Urbanik, K. (1972b). Slowly varying sequences of random variables. *Bulletin L'Académie Polonaise des Science, Série des Sciences Mathématiques, Astronomiques et Physiques, 20*, 679–682.
130. Urbanik, K. (1973). Limit laws for sequences of normed sums satisfying some stability conditions. In P. R. Krishnaiah (Ed.) *Multivariate analysis-III* (pp. 225–237). New York: Academic Press.
131. Vares, M. E. (1983). Local times for two-parameter Lévy processes. *Stochastic Processes and their Applications, 15*, 59–82.
132. Watanabe, T. (1996). Sample function behavior of increasing processes of class *L*. *Probability Theory and Related Fields, 104*, 349–374.
133. Watanabe, T. (1998). Sato's conjecture on recurrence conditions for multidimensional processes of Ornstein–Uhlenbeck type. *Journal of the Mathematical Society of Japan, 50*, 155–168.
134. Watanabe, T. (1999). On Bessel transforms of multimodal increasing Lévy processes. *Japanese Journal of Mathematics, 25*, 227–256.
135. Watanabe, T. (2000). Absolute continuity of some semi-selfdecomposable distributions and self-similar measures. *Probability Theory and Related Fields, 117*, 387–405.
136. Watanabe, T. (2001). Temporal change in distributional properties of Lévy processes. In O. E. Barndorff-Nielsen, T. Mikosch & S. I. Resnick (Eds.), *Lévy processes, theory and applications* (pp. 89–107). Boston: Birkhäuser.

137. Watanabe, T., & Yamamuro, K. (2010). Limsup behaviors of multi-dimensional selfsimilar processes with independent increments. *ALEA Latin American Journal of Probability and Mathematical Statistics, 7*, 79–116.
138. Widder, D. V. (1946). *The Laplace transform*. Princeton, NJ: Princeton University Press.
139. Wolfe, S. J. (1982a). On a continuous analogue of the stochastic difference equation $X_n = \rho X_{n-1} + B_n$. *Stochastic Processes and their Applications, 12*, 301–312.
140. Wolfe, S. J. (1982b). A characterization of certain stochastic integrals (Tenth Conference on Stochastic Processes and Their Applications, Contributed Papers). *Stochastic Processes and their Applications, 12*, 136.
141. Yamamuro, K. (2000a). Transience conditions for self-similar additive processes. *Journal of the Mathematical Society of Japan, 52*, 343–362.
142. Yamamuro, K. (2000b). On recurrence for self-similar additive processes. *Kodai Mathematical Journal, 23*, 234–241.
143. Yamazato, M. (1978). Unimodality of infinitely divisible distribution functions of class L. *Annals of Probability, 6*, 523–531.

Notation

$\mathbb{R}$, $\mathbb{Q}$, $\mathbb{Z}$, $\mathbb{N}$, and $\mathbb{C}$ are the sets of real numbers, rational numbers, integers, positive integers, and complex numbers, respectively. $\mathbb{R}_+ = [0, \infty)$ and $\mathbb{Z}_+ = \{0, 1, 2, \dots\}$.

$\mathbb{R}^d$ is the d-dimensional Euclidean space and elements of $\mathbb{R}^d$ are column vectors $x = (x_j)_{1 \le j \le d}$. The inner product is $\langle x, y \rangle = \sum_{j=1}^d x_j y_j$ for $x = (x_j)_{1 \le j \le d}$ and $y = (y_j)_{1 \le j \le d}$. The norm is $|x| = \langle x, x \rangle^{1/2}$. $S = \{\xi \in \mathbb{R}^d \colon |\xi| = 1\}$ is the unit sphere in $\mathbb{R}^d$.

$\mathbb{C}^N$ is the set of N-tuples of complex numbers and elements of $\mathbb{C}^N$ are column vectors $w = (w_j)_{1 \le j \le N}$. For any $w = (w_j)_{1 \le j \le N}$ and $v = (v_j)_{1 \le j \le N}$ in $\mathbb{C}^N$, we define $\langle w, v \rangle = \sum_{j=1}^N w_j v_j$. This is not the Hermitian inner product. For any $N' \times N$ matrix F, the transpose of F is denoted by $F^\top$. Thus $(w_1, \dots, w_N)^\top$ denotes the transpose of the row vector $(w_1, \dots, w_N)$.

For any Borel set C in $\mathbb{R}^d$, $\mathcal{B}(C)$ is the class of Borel subsets of C.

For any subset C of $\mathbb{R}^d$, $1_C(x)$ is the indicator function of C.

$\mathfrak{P} = \mathfrak{P}(\mathbb{R}^d)$ is the class of probability measures (distributions) on $\mathbb{R}^d$. That is, $\mu \in \mathfrak{P}(\mathbb{R}^d)$ is a countably additive mapping from $\mathcal{B}(\mathbb{R}^d)$ into $[0, 1]$ satisfying $\mu(\mathbb{R}^d) = 1$. $ID = ID(\mathbb{R}^d)$ is the class of infinitely divisible distributions on $\mathbb{R}^d$. $\mathfrak{S} = \mathfrak{S}(\mathbb{R}^d)$ is the class of stable distributions on $\mathbb{R}^d$. $\mathfrak{S}_\alpha = \mathfrak{S}_\alpha(\mathbb{R}^d)$ is the class of α-stable distributions on $\mathbb{R}^d$.

$\widehat{\mu}(z) = \int_{\mathbb{R}^d} e^{i\langle z, x \rangle} \mu(dx)$, $z \in \mathbb{R}^d$, is the *characteristic function* of $\mu \in \mathfrak{P}(\mathbb{R}^d)$.

$(\mu_1 * \mu_2)(B) = \int\int_{\mathbb{R}^d \times \mathbb{R}^d} 1_B(x + y) \mu_1(dx) \mu_2(dy)$, for $B \in \mathcal{B}(\mathbb{R}^d)$, is the *convolution* of μ_1 and μ_2 in $\mathfrak{P}(\mathbb{R}^d)$. For $\mu \in \mathfrak{P}$ and $n \in \mathbb{N}$, μ^n is the *n-fold convolution* of μ. For $\mu \in ID$ and $t \in \mathbb{R}_+$, μ^t is defined as $\widehat{\mu^t}(z) = e^{t(\log \widehat{\mu})(z)}$, where $\log \widehat{\mu}$ is the *distinguished logarithm* of $\widehat{\mu}$. Sometimes μ^n and μ^t are written as μ^{n*} and μ^{t*} , respectively.

$\mathcal{L}(X)$ is the distribution (law) of a random variable X on $\mathbb{R}^d$. $X \overset{d}{=} Y$ means that two random variables X and Y have a common distribution or they are *identical in law*, that is, $\mathcal{L}(X) = \mathcal{L}(Y)$. $\{X_t\} \overset{d}{=} \{Y_t\}$ means that two stochastic processes $\{X_t\}$

A. Rocha-Arteaga, K. Sato, *Topics in Infinitely Divisible Distributions and Lévy Processes*, SpringerBriefs in Probability and Mathematical Statistics,
https://doi.org/10.1007/978-3-030-22700-5

and $\{Y_t\}$ are *identical in law*, that is, have a common system of finite-dimensional distributions. Note that $X_t \stackrel{d}{=} Y_t$ simply means that, for each t, X_t and Y_t have a common distribution.

For any $\mu \in \mathfrak{P}(\mathbb{R}^d)$ with $\int |x|^2 \mu(dx) < \infty$, the *covariance matrix* of μ is the nonnegative-definite matrix $(\mathrm{cov}(X_i, X_j))_{i,j=1}^d$ for a random variable $(X_j)_{1 \le j \le d}$ such that $\mathcal{L}((X_j)_{1 \le j \le d}) = \mu$.

For μ_n ($n = 1, 2, \ldots$) and μ in $\mathfrak{P}$, $\mu_n \to \mu$ means *weak convergence* of μ_n to μ, that is, $\int f(x)\mu_n(dx) \to \int f(x)\mu(dx)$ for all bounded continuous functions f. When μ_n ($n = 1, 2, \ldots$) and μ are finite measures on $\mathbb{R}^d$, the convergence $\mu_n \to \mu$ is defined in the same way.

δ_c is a distribution concentrated at c; it is called a *trivial* distribution. A random variable X is *trivial* if $\mathcal{L}(X)$ is trivial. A stochastic process $\{X_t\}$ is a *trivial* process if X_t is trivial for all t; it is a *zero process* if $\mathcal{L}(X_t) = \delta_0$ for all $t \ge 0$.

The words *increasing* and *decreasing* are used in the wide sense allowing flatness.

Unless specifically mentioned, *measurable* means Borel measurable.

Index

A. Rocha-Arteaga, K. Sato, *Topics in Infinitely Divisible Distributions and Lévy Processes*, SpringerBriefs in Probability and Mathematical Statistics,
https://doi.org/10.1007/978-3-030-22700-5

Printed in the United States
By Bookmasters